LANDSCAPE ECOLOGY PRINCIPLES

in LANDSCAPE ARCHITECTURE

and LAND-USE PLANNING

Wenche E. Dramstad, James D. Olson, and Richard T. T. Forman

Harvard University
Graduate School of Design

Island Press

American Society of Landscape Architects

Library of Congress Catalog Card Number 95-82343

International Standard Book Number 1-55963-514-2

Published by Harvard University Graduate School of Design, Island Press and the American Society of Landscape Architects

Island Press, 1718 Connecticut Avenue, N.W., Suite 300, Washington, DC 20009

Printed on recycled acid-free paper

Cover Photos:
Idaho, U.S.A., USDA Soil Conservation Service photo.
California, U.S.A., USDA Soil Conservation Service photo.
Massachusetts, U.S.A., R. Forman photo.

Cover Illustration:
James D. Olson and Richard T.T. Forman

CONTENTS

FOREWORD

The thin mosaic, the tissue of the planet, is in upheaval. An urgent need exists for new tools and new language to understand how to live without losing nature. The solutions will be at the landscape scale—working with the larger pattern, understanding how it works, and designing in harmony with the structure of the natural system that sustains us all.

Each landscape has its own signature. This book will give you new eyes, and a means to communicate and collaborate with the many ecologists and landscape architects who are reaching out to work together and find cross-disciplinary solutions to land-use challenges.

Places are like large "organisms," the products of natural forms and processes at work. Places are uniquely different and each possesses an intrinsic potential for change. This book will also help landscape architects and planners to work with communities that are inventing and formulating the new civics of sustainability.

What encourages me most about this book is how its principles are both simple and holistic in the way they tie together land, water, wildlife, and people. As designers and planners we must weave together this mosaic of patches and corridor networks, like a quilt held together with threads, to hold the landscape from falling apart. Understanding this mosaic will be our greatest challenge.

We need more succinct books like this one, with its simple tools and language, to couple the usually opposing forces of government regulations, economic self-interest, and the land ethic to run parallel.

Grant Jones, FASLA
Jones & Jones
Seattle, Washington

PREFACE AND ACKNOWLEDGMENTS

Landscape ecology has rapidly emerged in the past decade to become usable and important to practicing land-use planners and landscape architects. The focus on heterogeneous land mosaics, such as neighborhoods, whole landscapes, and regions, is at increasingly the critical spatial scale. Animals, plants, water, materials, and energy are spatially distributed, move, flow, and change in predictable ways in these mosaics. Thus professionals and scholars have incorporated bits of the new field in their work. But many have also requested a summary of key principles, and how they might be applied in design and planning.

This publication is therefore a handbook or primer, listing and illustrating many key principles. It also provides examples of how the principles can be applied in design, planning, and solving vexing land-use issues.

The book is not a cookbook giving exact ingredients and steps. Designers and planners are rife with creativity and original ideas. The principles presented are solid background colors on the professional's palette, the foundations that are combined to produce important new designs and solutions.

If society decides, for example, to add a road, a nature reserve, or a housing tract, these principles will help accomplish the goal by maximizing ecological integrity, and minimizing land degradation. Furthermore, principles at this relatively broad scale become a surrogate for long time. They nudge society into long-term planning and decision-making.

Using the principles is not difficult, and leads to more integrative designs and plans. It helps reduce the landscape fragmentation and degradation so evident around us. Individual professionals familiar with landscape ecology already accomplish these specific results.

In addition though, solutions to environmental and societal problems require cross-disciplinary design and planning by groups. Another objective of this book is to strengthen the two-way street between ecologists and planners/landscape architects. Plenty of ecologists will also read this book, and some will take a deeper interest in landscape architecture and land-use planning. Such a synergism will result in deeper understanding of the landscape ecology principles, the development of additional useful principles, and their better application in land planning and design.

We are pleased to acknowledge the key financial support of the following organizations that made this book possible:

Agricultural University of Norway

Harvard University Graduate School of Design

The Research Council of Norway

Sasaki Associates, Inc.

We also deeply appreciate and are delighted to acknowledge the following persons: J. Thomas Atkins (Jones & Jones, Seattle, Washington), Margot D. Cantwell (Environmental Design and Management, Halifax, Canada), Leslie Kerr (U. S. Fish and Wildlife Service, Anchorage, Alaska), Alistair T. McIntosh (Sasaki Associates, Watertown, Massachusetts), and Mary Ann Thompson (Thompson and Rose Architects, Cambridge, Massachusetts) provided important critical reviews from the perspective of practicing professionals. Carl Steinitz (Harvard University) kindly permitted us to test an earlier draft in his class studio. Tricia Bales, Jorgen Blomberg, Jennifer Brooke, Mona Campbell, Lisa Cloutier, Mark Daley, Edie Drcar, Robert Hopper, Frank Kluber, Francesca Levaggi, Justine Lovinger, Haruko Masutani, Koa Pickering, Hillary Quarles, Aya Sakai, Carrie Steinbaum, and Lital Szmuk provided very useful comments, based on using the book draft during an academic landscape-planning project at Harvard. And Gareth L. A. Fry (Norwegian Institute for Nature Research), Jan Heggenes (Department of Biology and Nature Conservation, Agricultural University of Norway), Sharon K. Collinge (University of California–Davis), J. Douglas Olson, Davorin Gazvoda, Rodney Hoinckes, and Michael W. Binford (Harvard University) offered much valuable advice and support.

Wenche E. Dramstad[1]
James D. Olson[2]
Professor Richard T. T. Forman
Harvard University
Graduate School of Design

Present Addresses:

[1] Agricultural University of Norway
Department of Biology and Nature Conservation
P.O. Box 5014
N-1432 Ås, Norway

[2] 4 Tamworth Road
Waban, MA 02168

Biodiversity must be conserved as a matter of principle, as a matter of survival, and as a matter of economic benefit.

UNEP, IUCN and WWF in their joint report, *Caring for the Earth*, 1992

Time changes

History indicates that in the face of crisis, human ingenuity, creativity, discoveries, inventions, and new solutions cascade forth. Today almost all major studies point to a coalescence in the next few decades of significant land degradation, population growth, water shortage, fertile soil erosion, biodiversity loss, and spread of huge urban areas. Society is comfortable in thinking of small spaces and short times, or at best considering trends separately. When the trends are connected, it is hard to miss the crisis looming. The timetable says we and our children will be there. At center stage will be land-use pattern.

Wild turkeys crossing an opening, Texas, U.S.A., USDA Soil Conservation Service photo.

Land planners and landscape architects are uniquely poised to play key roles for society, to provide new solutions. These are professionals and scholars who focus on the land. Solve problems. Design and create plans. Look to the future. Are optimists, can-do people. Are synthesizers who weave diverse needs together into a whole. Have ingenuity and creativity. Know aesthetics or economics. Know that human culture is essential in a design or plan. And know that ecological integrity of the land is critical.

Gold mining spoils on floodplain, Montana, U.S.A., USDA Soil Conservation Service photo.

Landscape architecture and land-use planning have a long and distinguished history of inspired accomplishments. The images of extensive Italian country villas, 19th-century planning and design of major American cities, and the 20th-century development of national parks are impressive harmonies in the land. A key to their brilliance is the enlightened meshing of nature and culture.

The designers and planners were not amateurs in either nature or culture, but had extensive education and knowledge in both. *Nature* included the biological patterns and physical processes entwined in vegetation, wildlife populations, species richness, wind, water, wetlands, and aquatic communities. *Culture* integrated the diverse human dimensions of economics, aesthetics, community social patterns, recreation, transportation, and sewage/waste handling.

Channelized stream corridor, Georgia, U.S.A., USDA Soil Conservation Service photo.

What are the natural features which make a township handsome? A river, with its waterfalls and meadows, a lake, a hill, a cliff or individual rocks, a forest, and ancient trees standing singly. Such things are beautiful; they have a high use which dollars and cents never represent. If the inhabitants of a town were wise, they would seek to preserve these things, though at a considerable expense; for such things educate far more than any hired teachers or preachers, or any present recognized system of school education. I do not think him fit to be the founder of a state or even of a town who does not foresee the use of these things....

Henry David Thoreau, *Journal*, 1861

In some countries these two basic components—ecology and culture—have diverged relatively recently. For example, ecology has matured, and veered away from planning and design. Or economics has become paramount. Or aesthetics. Or sewage and wastes have been considered only an engineering problem. Or flourishing litigation has colored decision-making. Or local actions have overridden regional thinking and planning. These sound so familiar to professionals in the field. The deeper message is the importance of a new form of linkage between ecology and culture, land and people, nature and humans.

There is an increasing evidence suggesting that mental health and emotional stability of populations may be profoundly influenced by frustrating aspects of an urban, biologically artificial environment. It seems likely that we are genetically programmed to a natural habitat of clean air and a varied green landscape, like any other mammal. The specific physiological reactions to natural beauty and diversity, to the shapes and colors of nature, especially to green, to the motions and sounds of other animals, we do not comprehend and are reluctant to include in studies of environmental quality. Yet it is evident that in our daily lives nature must be thought of not as a luxury to be made available if possible, but as part of our inherent indispensable biological need.

Frederick Law Olmsted, in *Biography*, by J.E. Todd, 1982.

The missing ingredient and key to the new weaving appeared in the 1980s, and mushroomed in the 1990s. Landscape ecology, the ecology of large heterogeneous areas, of landscapes, of regions, of portions thereof, or simply of land mosaics, has increasingly appeared on the palette.

It is at exactly the right spatial scale. It explicitly integrates nature and humans. Its principles work in any landscape, from urban to pastureland and desert to tundra. Its spatial language is simple, catalyzing ready communication among land-use decision-makers, professionals, and scholars of many disciplines. And it is no academic musing, but, centered on spatial pattern, is easily and directly usable. It often evokes, "Why didn't we think of that?" or "Good to know there's science behind it now."

Urban-rural edge, Virginia, U.S.A., USDA Soil Conservation Service photo.

Landscape architects and land-use planners will always be experts in small areas, the tiny parks, housing clusters, and shopping malls. At the same time, such professionals also know that only designing and planning little pieces of the land leads to a fragmented world that doesn't work, either ecologically or for people. Fortunately, the knowledgeable meshing of humans and ecology at a broader scale is now in the repertoire, and will become routine. The solution for a small economically or aesthetically focused project will emanate as much from the surrounding mosaic pattern as from the site itself. And the larger land-area project will focus directly on spatial pattern, movements, and changes of its mosaic, based solidly on principles of landscape and regional ecology.

Objectives

The objectives of this book are to: 1. Pinpoint many key principles of landscape ecology, especially those directly usable in land-use planning and landscape architecture. 2. Illustrate how principles can be used in planning and design projects.

Development of landscape ecology

The key literature and concepts of landscape architecture and land-use planning are doubtless well known to the reader. However, a brief background in landscape ecology appears useful. The foundations may be traced back to scholars up to about 1950, who elucidated the natural history and physical environment patterns of large areas. Certain geographers, plant geographers, soil scientists, climatologists, and natural history writers were the "giants with shoulders" upon which later work stood.

From about 1950 to 1980 diverse important threads emerged, and their weaving together commenced. The term landscape ecology was used when aerial photography began to be widely available. The concept focused on specific spatial pattern in a section of a landscape, where biological communities interacted with the physical environment (Troll 1939, 1968). Diverse definitions of the term of course have appeared over the years, but today the primary, most widely held concept is as follows.

Ecology is generally defined as the study of the interactions among organisms and their environment, and a *landscape* is a kilometers-wide mosaic over which particular local ecosystems and land-uses recur. These concepts have proven to be both simple and operationally useful. Thus *landscape ecology* is simply the ecology of landscapes, and *regional ecology* the ecology of regions.

Fields, woods, and hedgerows, New Jersey, U.S.A., USDA Soil Conservation Service photo.

Several other disciplines or important concepts were incorporated during this weaving phase of landscape ecology. The ecosystem concept, animal and plant geography, vegetation methodology, hedgerow studies, agronomic studies, and island biogeographic theory were important. Also quantitative geography, regional studies, human culture and aesthetics, and land evaluation were incorporated. Landscape architecture and land-use planning literature began to be included. This phase produced an abundance of intriguing, interdisciplinary individual designs, but no clear form of the overall tapestry was evident.

Fields, wooded patches, and wooded corridors, England, R. Forman photo.

Since about 1980 the "land mosaic" phase has coalesced, where puzzle pieces increasingly fit together and an overall conceptual design of landscape and regional ecology emerges. Edited books tend to compile disparate, but sometimes key, pieces of landscape ecology. These include general concepts (Tjallingii & de Veer 1981, Ruzicka 1982, Brandt & Agger 1984, Zonneveld & Forman 1990), habitat fragmentation and conservation (Burgess & Sharpe 1981, Saunders et al. 1987, Hansson & Angelstam 1991), corridors and connectivity (Schreiber 1988, Brandle et al. 1988, Saunders & Hobbs 1991, Smith & Hellmund 1993), quantitative methodology (Berdoulay & Phipps 1985, Turner & Gardner 1991), and heterogeneity, boundaries, and restoration (Turner 1987, Hansen & di Castri 1992, Vos & Opdam 1992, Saunders et al. 1993).

The major authored volumes, in contrast, tend to integrate and synthesize theory and concepts. These books include land evaluation and planning (Zonneveld 1979, Takeuchi 1991), soil and agriculture (Vink 1980), logging and conservation (Harris 1984), total human ecosystem (Naveh & Lieberman 1993), hierarchy theory (O'Neill et al. 1986), statistical methodology (Jongman et al. 1987), river corridors (Malanson 1993), and land mosaics (Forman & Godron 1986, Forman 1995). Of course, to gain a solid and full understanding of the subject, articles in *Landscape Ecology* and many other journals are a must, and often a delight.

Forest clearcuts and logging roads, Oregon, U.S.A., R. Forman photo.

Landscape ecology today

The principles of landscape and regional ecology apply in any land mosaic, from suburban to agriculture and desert to forest. They work equally in pristine natural areas and areas of intense human activity. The object spread out beneath an airplane, or in an aerial photograph, contains living organisms in abundance, and therefore is a living system.

Like a plant cell or a human body, this living system exhibits three broad characteristics: structure, functioning, and change. *Landscape structure* is the spatial pattern or arrangement of landscape elements. *Functioning* is the movement and flows of animals, plants, water, wind, materials, and energy through the structure. And *change* is the dynamics or alteration in spatial pattern and functioning over time.

The structural pattern of a landscape or region is composed entirely of three types of elements. Indeed, these universal elements—patches, corridors, and matrix—are the handle for comparing highly dissimilar landscapes and for developing general principles. They also are the handle for land-use planning and landscape architecture, since spatial pattern strongly controls movements, flows, and changes.

The simple spatial language becomes evident when considering how patches, corridors, and the matrix combine to form the variety of land mosaics on earth. What are the key attributes of *patches*? They are large or small, round or elongated, smooth or convoluted, few or numerous, dispersed or clustered, and so forth. What about *corridors*? They appear narrow or wide, straight or curvy, continuous or disconnected, and so on. And the *matrix* is single or subdivided, variegated or nearly homogeneous, continuous or perforated, etc. These spatial attributes or descriptors are close to dictionary definitions, and all are familiar to decision-makers, professionals, and scholars of many disciplines.

The whole landscape or region is a mosaic, but the local neighborhood is likewise a configuration of patches, corridors, and matrix. Landscape ecologists are actively studying and developing principles for the biodiversity patterns and natural processes in these configurations or neighborhood mosaics.

For example, changing a mosaic by adding a hedgerow, pond, house, woods, road, or other element changes the functioning. Animals change their routes, water flows alter direction, erosion of soil particles changes, and humans move differently. Removing an element alters flows in a different manner. And rearranging the existing elements causes yet greater changes in how the neighborhood functions. These spatial elements and their arrangements are the ready handles for landscape architects and land-use planners.

Road corridor, Western Australia, Photo courtesy of B.M.J. (Penny) Hussey.

Natural processes as well as human activities change landscapes. In a time series of aerial photographs a sequence of mosaics typically appears. Habitat fragmentation is frequently noted and decried. But many other spatial processes are evident in land transformation, such as perforation, dissection, shrinkage, attrition, and coalescence, each with major ecological and human implications.

In short, the landscape ecology principles in this book are directly applicable and offer opportunities for wise planning, design, conservation, management, and land policy. The principles are significant from neighborhood to regional mosaics. They focus on spatial pattern, which strongly determines functioning and change. Their patch-corridor-matrix components have universality for any region. And their language enhances communication and collaboration. They will become central as society begins to seriously address the issue of creating sustainable environments.

Strips and pond made for wildlife, Texas, U.S.A., USDA Soil Conservation Service photo.

Roadmap

Part I presents the landscape ecological principles. For convenience these are grouped by patches, edges, corridors, and mosaics. Part II then illustrates practical applications of the principles. This begins with schematic applications at broad, medium, and fine scales. It ends with encapsulated case studies from around the world.

REFERENCES

Berdoulay, V. and M. Phipps, eds. 1985. *Paysage et Système*. Editions de l'Université d'Ottawa, Ottawa.

Brandle, J.R., D.L. Hintz and J.W. Sturrock, eds. 1988. *Windbreak Technology*. Elsevier, Amsterdam. (Reprinted from *Agriculture, Ecosystems and Environment* 22-23, 1988).

Brandt, J. and P. Agger, eds. 1984. *Proceedings of the First International Seminar on Methodology in Landscape Ecology Research and Planning*. 5 vols. Roskilde Universitetsforlag GeoRuc, Roskilde, Denmark.

Burgess, R.L. and D.M. Sharpe, eds. 1981. *Forest Island Dynamics in Man-dominated Landscapes*. Springer-Verlag, New York.

Forman, R.T.T., ed. 1979. *Pine Barrens: Ecosystem and Landscape*. Academic Press, New York.

Forman, R.T.T. 1995. *Land Mosaics: The Ecology of Landscapes and Regions*. Cambridge University Press, Cambridge.

Forman, R.T.T. 1995. Some general principles of landscape and regional ecology. *Landscape Ecology* 10: 133-142.

Forman, R.T.T. and M. Godron. 1986. *Landscape Ecology*. John Wiley, New York.

Hansen, A.J. and F. di Castri, eds. 1992. *Landscape Boundaries: Consequences for Biotic Diversity and Ecological Flows*. Springer-Verlag, New York.

Hansson, L. and P. Angelstam. 1991. Landscape ecology as a theoretical basis for nature conservation. *Landscape Ecology* 5: 191-201.

Harris, L.D. 1984. *The Fragmented Forest: Island Biogeography Theory and the Preservation of Biotic Diversity*. University of Chicago Press, Chicago.

Hobbs, R.J. 1995. Landscape ecology. *Encyclopedia of Environmental Biology* 2, pp. 417-428.

Jongman, R.G.H., C.J.F. ter Braak and O.F.R. van Tongeren. 1987. *Data Analysis in Community and Landscape Ecology*. PUDOC, Wageningen, Netherlands.

Malanson, G.P. 1993. *Riparian Landscapes*. Cambridge University Press, Cambridge.

Naveh, Z. and A.S. Lieberman. 1993. *Landscape Ecology: Theory and Application*. Springer-Verlag, New York.

O'Neill, R.V., D.L. DeAngelis, J.B. Waide and T.F.H. Allen. 1986. *A Hierarchical Concept of Ecosystems*. Princeton University Press, Princeton.

Ruzicka, M., ed. 1982. *Proceedings of the VIth International Symposium on Problems in Landscape Ecological Research*. Institute for Experimental Biology and Ecology, Bratislava, Czechoslovakia.

Saunders, D.A., G.W. Arnold, A.A. Burbidge and A.J.M. Hopkins, eds. 1987. *Nature Conservation: The Role of Remnants of Native Vegetation*. Surrey Beatty, Chipping Norton, Australia.

Saunders, D.A. and R.J. Hobbs, eds. 1991. *Nature Conservation 2: The Role of Corridors*. Surrey Beatty, Chipping Norton, Australia.

Saunders, D.A., R.J. Hobbs and P.R. Ehrlich, eds. 1993. *Nature Conservation 3: The Reconstruction of Fragmented Ecosystems: Global and Regional Perspectives*. Surrey Beatty, Chipping Norton, Australia.

Schreiber, K-F. 1988. *Connectivity in Landscape Ecology*. Münstersche Geographische Arbeiten 29, Ferdinand Schoningh, Paderborn, Germany.

Smith, D.S. and P.C. Hellmund, eds. 1993. *Ecology of Greenways: Design and Function of Linear Conservation Areas*. University of Minnesota Press, Minneapolis, Minnesota.

Takeuchi, K. 1991. *Regional (Landscape) Ecology*. (In Japanese). Asakura Publishing, Tokyo.

Tjallingii, S.P. and A.A. de Veer, eds. 1981. *Perspectives in Landscape Ecology*. PUDOC, Wageningen, Netherlands.

Torrey, B. and F.H. Allen. 1962. *The Journal of Henry D. Thoreau*. 14 vols. Dover Publications, New York.

Troll, C. 1939. Luftbildplan und ökologische Bodenforschung. *Zeitschrift der Gesellschaft für Erdkunde zu Berlin*, pp. 241-298.

Troll, C. 1968. Landschaftsokologie. In Tuxen, R., ed. *Pflanzensoziologie und Landschaftsokologie*, pp. 1-21. Dr. W. Junk Publishers, The Hague, Netherlands.

Turner, M.G., ed. 1987. *Landscape Heterogeneity and Disturbance*. Springer-Verlag, New York.

Turner, M.G. 1989. Landscape ecology: the effect of pattern on process. *Annual Review of Ecology and Systematics* 20, pp. 171-197.

Turner, M.G. and R.H. Gardner, eds. 1991. *Quantitative Methods in Landscape Ecology: The Analysis and Interpretation of Landscape Heterogeneity*. Springer-Verlag, New York.

Vink, A.P.A. 1980. *Landschapsecologie en Landgebruik*. Bohn, Scheltema and Holkema, Utrecht, Netherlands. (1983 translation. Landscape Ecology and Land Use. Longman, London).

Vos, C.C. and P. Opdam, eds. 1992. *Landscape Ecology of a Stressed Environment*. Chapman and Hall, London.

Zonneveld, I.S. 1979. *Land Evaluation and Land(scape) Science*. 2nd edition. ITC Textbook VII.4. International Institute for Aerial Survey and Earth Sciences, Enschede, Netherlands.

Zonneveld, I. S. and R. T. T. Forman, eds. 1990. *Changing Landscapes: An Ecological Perspective*. Springer-Verlag, New York.

PATCHES

Landscape ecology principles are listed and illustrated below in four sections: Patches; Edges; Corridors; and Mosaics. Each section begins with an introduction to important terms and concepts, and ends with a list of key references. For additional references please refer to the bibliography.

In a densely populated world plant and animal habitat increasingly appears in scattered patches. Ecologists first considered habitat patches analogous with islands, but soon largely abandoned the analogy due to the major differences between the sea and the matrix of countryside and suburban developments surrounding a "terrestrial" patch. Patches, however, do exhibit a degree of isolation, the effect and severity being dependent on the species present.

Farmstead woodlots and wheat, Minnesota, U.S.A., USDA Soil Conservation Service photo.

Four *origins* or causes of vegetation patches are usefully recognized: *remnants* (e.g., areas remaining from an earlier more extensive type, such as woodlots in agricultural areas); *introduced* (e.g., a new suburban development in an agricultural area, or a small pasture within a forest); *disturbance* (e.g., a burned area in a forest, or a spot devastated by a severe windstorm); and *environmental resources* (e.g., wetlands in a city, or oases in a desert).

Patches are analyzed below and differentiated in terms of (1) size, (2) number, and (3) location. Patches may be as *large* as a national forest, or as *small* as a single tree. Patches may be *numerous* in a landscape, such as avalanches or rock slides on a mountainside, or be *scarce* such as oases in a desert. The location of patches may be *beneficial* or *deleterious* to the optimal functioning of a landscape. For example, small, remnant forest patches between large reserves in an agricultural matrix can be beneficial. In contrast, a landfill located adjacent to a sensitive wetland may have a negative impact on the ecological health of the landscape.

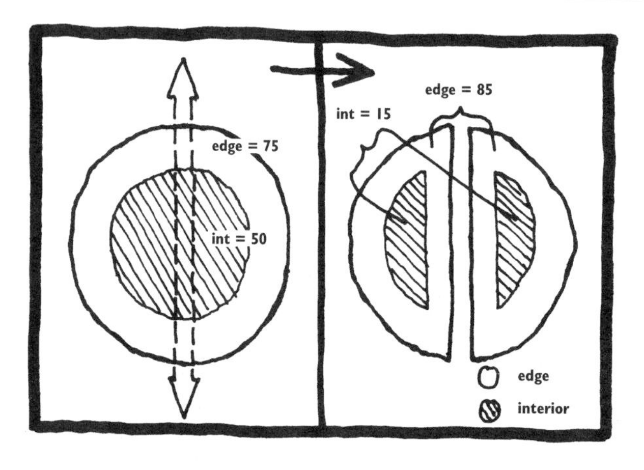

P1. Edge habitat and species

Dividing a large patch into two smaller ones creates additional edge habitat, leading to higher population sizes and a slightly greater number of edge species, which are often common or widespread in the landscape.

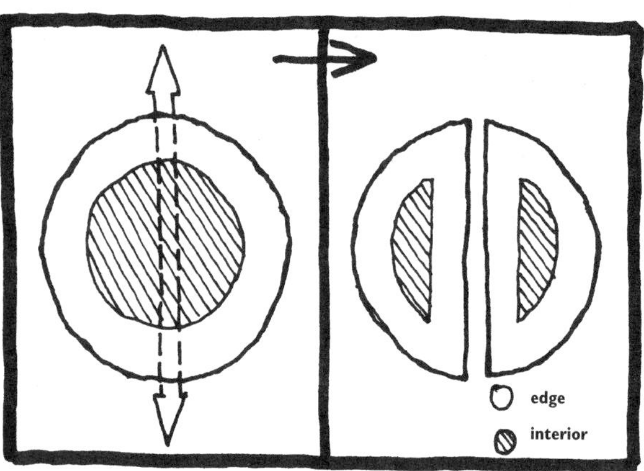

P2. Interior habitat and species

Dividing a large patch into two smaller ones removes interior habitat, leading to reduced population sizes and number of interior species, which are often of conservation importance.

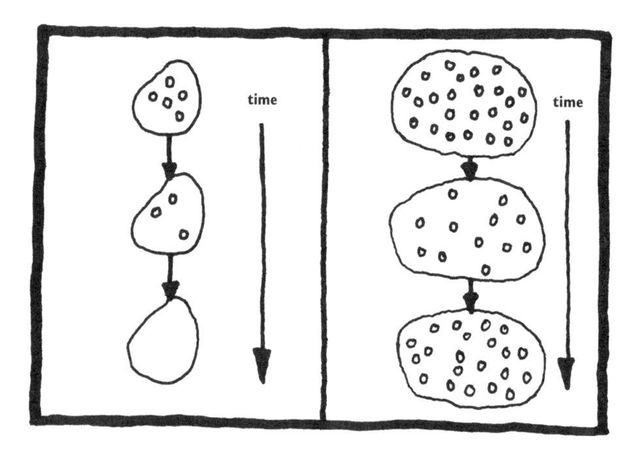

P3. Local extinction probability

A larger patch normally has a larger population size for a given species than a smaller patch, making it less likely that the species (which fluctuates in population size) will go locally extinct in the larger patch.

P4. Extinction

The probability of a species becoming locally extinct is greater if a patch is small, or of low habitat quality.

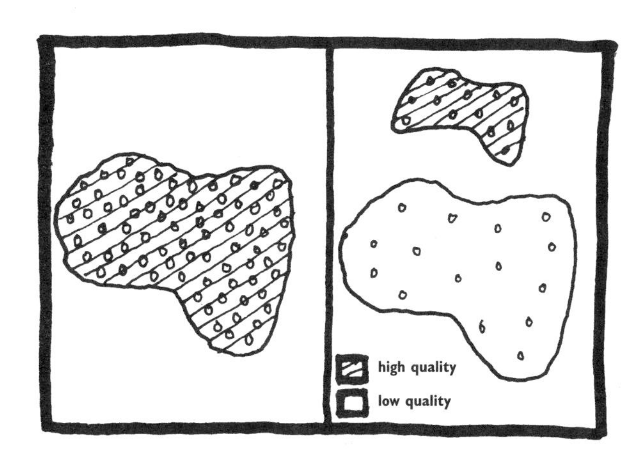

P5. Habitat diversity

A large patch is likely to have more habitats present, and therefore contain a greater number of species than a small patch.

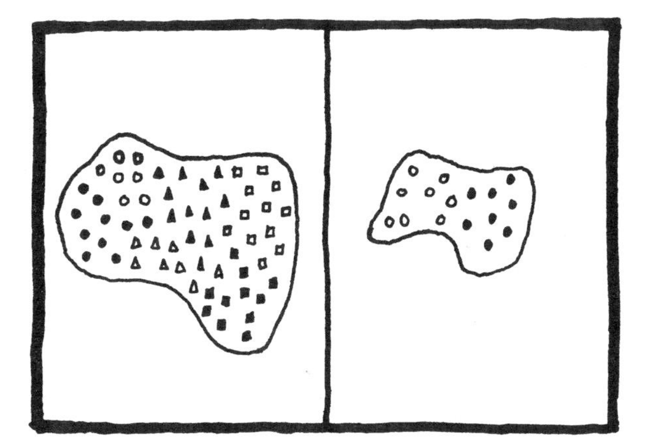

P6. Barrier to disturbance

Dividing a large patch into two smaller ones creates a barrier to the spread of some disturbances.

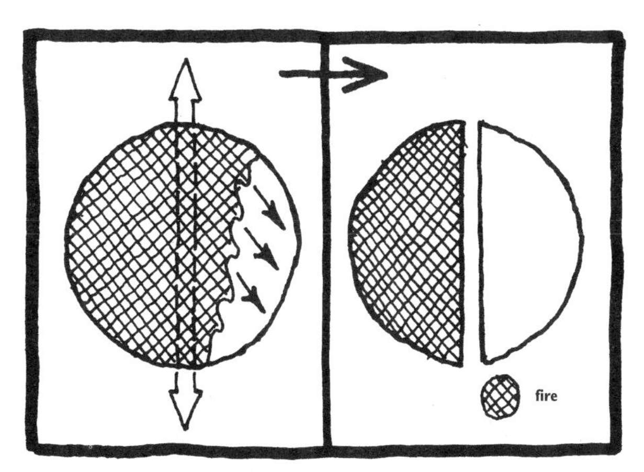

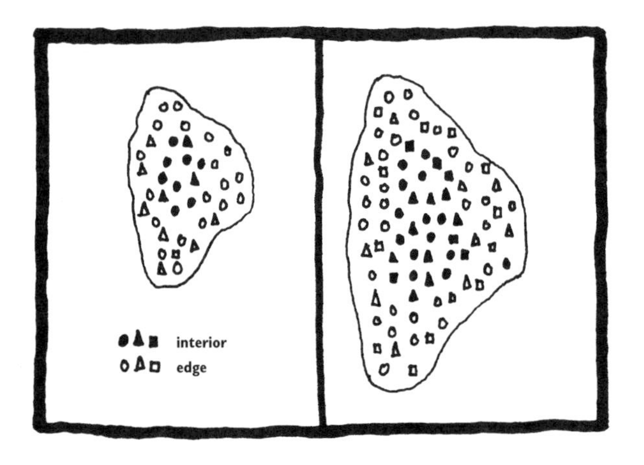

P7. Large patch benefits

Large patches of natural vegetation are the only structures in a landscape that protect aquifers and interconnected stream networks, sustain viable populations of most interior species, provide core habitat and escape cover for most large-home-range vertebrates, and permit near-natural disturbance regimes.

P8. Small patch benefits

Small patches that interrupt extensive stretches of matrix act as stepping stones for species movement. They also contain some uncommon species where large patches are absent or, in unusual cases, are unsuitable for a species. Therefore small patches provide different and supplemental ecological benefits than large patches.

PATCH NUMBER: HOW MANY?

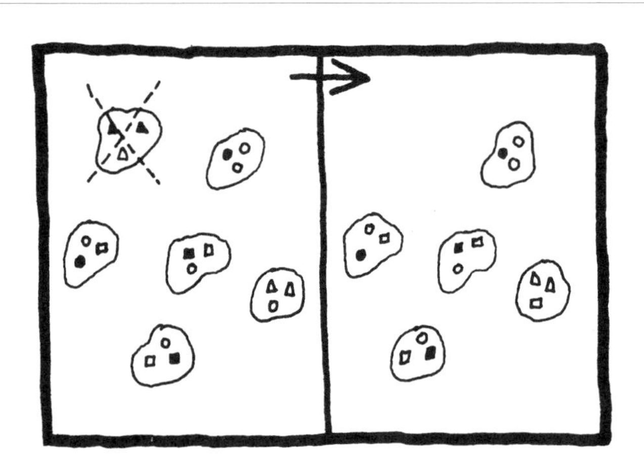

P9. Habitat loss

Removal of a patch causes habitat loss, which often reduces the population size of a species dependent upon that habitat type, and may also reduce habitat diversity, leading to fewer species.

P10. Metapopulation dynamics

Removal of a patch reduces the size of a metapopulation (i.e., an interacting population subdivided among different patches), thereby increasing the probability of local within-patch extinctions, slowing down the recolonization process, and reducing stability of the meta-population.

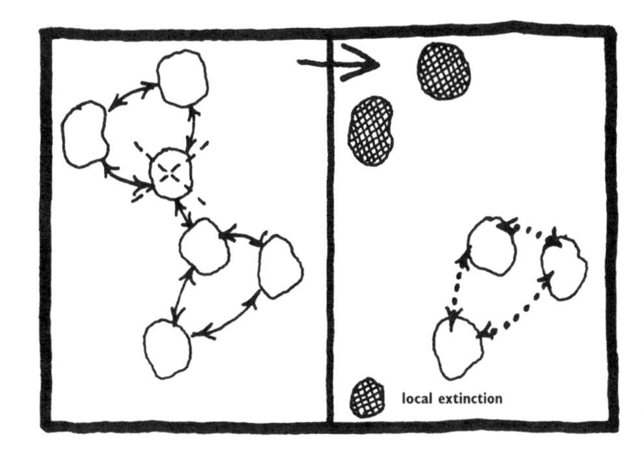

P11. Number of large patches

Where one large patch contains almost all the species for that patch type in the landscape, two large patches may be considered the minimum for maintaining species richness. However, where one patch contains a limited portion of the species pool, up to four or five large patches are probably required.

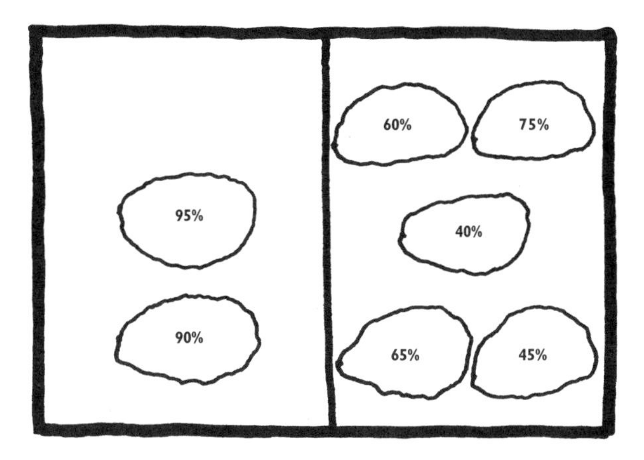

P12. Grouped patches as habitat

Some relatively generalist species can, in the absence of a large patch, survive in a number of nearby smaller patches, which although individually inadequate, are together suitable.

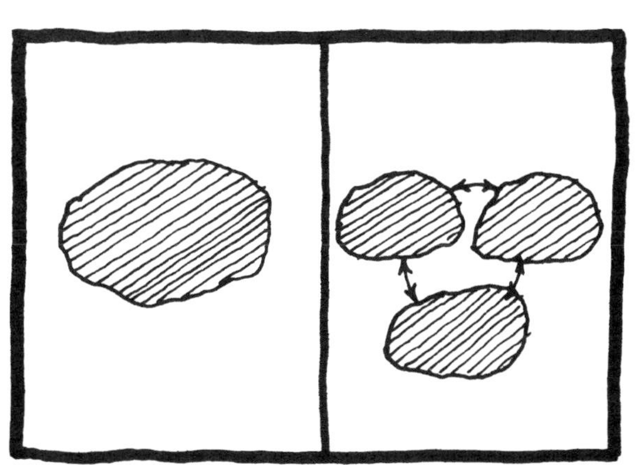

P13. Extinction

The probability of a species going locally extinct is greater in an isolated patch. Isolation is a function not only of distance, but also of the characteristics (i.e., resistance) of the intervening matrix habitat.

P14. Recolonization

A patch located in close proximity to other patches or the "mainland" will have a higher chance of being (re-)colonized within a time interval, than a more isolated patch.

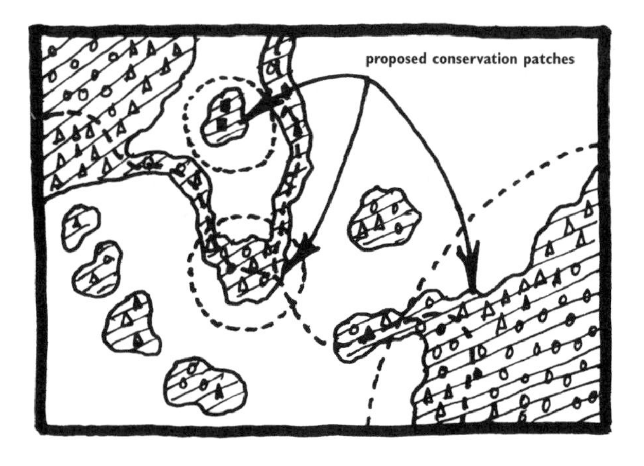

P15. Patch selection for conservation

The selection of patches for conservation should be based on their: *1) contribution to the overall system*, i.e., how well the location of a patch relates or links to other patches within the landscape or region; and *2) unusual or distinctive characteristics*, e.g., whether a patch has any rare, threatened, or endemic species present.

KEY REFERENCES

Forman, R.T.T. 1995. *Land Mosaics: The Ecology of Landscapes and Regions*. Cambridge University Press, Cambridge. [Patch size, number, and location]

Forman, R.T.T., A.E. Galli, and C.F. Leck. 1976. "Forest size and avian diversity in New Jersey woodlots with some land use implications." *Oecologia* 26, pp. 1-8. [Patch number]

Game, M., and G.F. Peterken. 1984. "Nature reserve selection strategies in the woodlands of Central Lincolnshire, England." *Biological Conservation* 29, pp. 157-181. [Patch number]

Harris, L.D. 1984. *The Fragmented Forest: Island Biogeography Theory and the Preservation of Biotic Diversity*. University of Chicago Press, Chicago. [Patch size and location]

Opdam, P. 1991. "Metapopulation theory and habitat fragmentation: a review of Holarctic breeding bird studies." *Landscape Ecology* 5, pp. 93-106. [Patch size]

Saunders, D.A., G.W. Arnold, A.A. Burbidge, and A.J.M. Hopkins, eds. 1987. *Nature Conservation: The Role of Remnants of Native Vegetation*, Surrey Beatty, Chipping Norton, Australia. [Patch size and location]

Shafer, C.L. 1990. *Nature Reserves: Island Theory and Conservation Practice*. Smithsonian Institution Press, Washington, D.C. [Patch size, number, and location]

van Dorp, D. and P.F.M. Opdam. 1987. "Effects of patch size, isolation and regional abundance on forest bird communities." *Landscape Ecology* 1, pp. 59-73. [Patch location]

See additional references on page 71

EDGES AND BOUNDARIES

An *edge* is described as the outer portion of a patch where the environment differs significantly from the interior of the patch. Often, edge and interior environments simply look and feel differently. For example, vertical and horizontal structure, width, and species composition and abundance, in the edge of a patch, differ from interior conditions, and together comprise the *edge effect*. Whether a boundary is curvilinear or straight influences the flow of nutrients, water, energy, or species along or across it.

Convoluted grassland-forest boundary, Idaho, U.S.A., R. Forman photo.

Boundaries may also be "political" or "administrative," that is artificial divisions between inside and out, which may or may not correspond to natural "ecological" boundaries or edges. Relating these artificial edges with natural ones is important. As human development continues its expansion into natural environments, the edges created will increasingly form the critical point for interactions between human-made and natural habitats.

The shapes of patches, as defined by their boundaries, can be manipulated by landscape architects and land-use planners to accomplish an ecological function or objective. Due to the diverse significance of edges, rich opportunities exist to use this key ecological transition zone between two types of habitat in designs and plans.

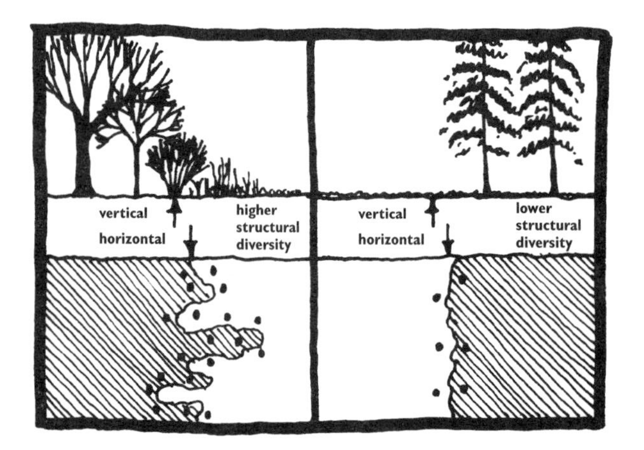

E1. Edge structural diversity

Vegetative edges with a high structural diversity, vertically or horizontally, are richer in edge animal species.

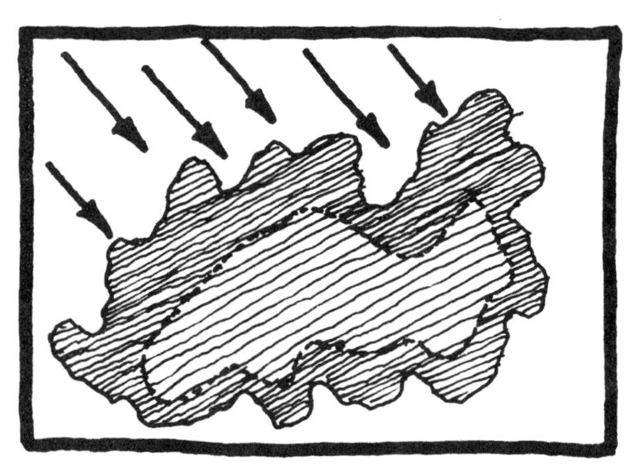

E2. Edge width

Edge width differs around a patch, with wider edges on sides facing the predominant wind direction and solar exposure.

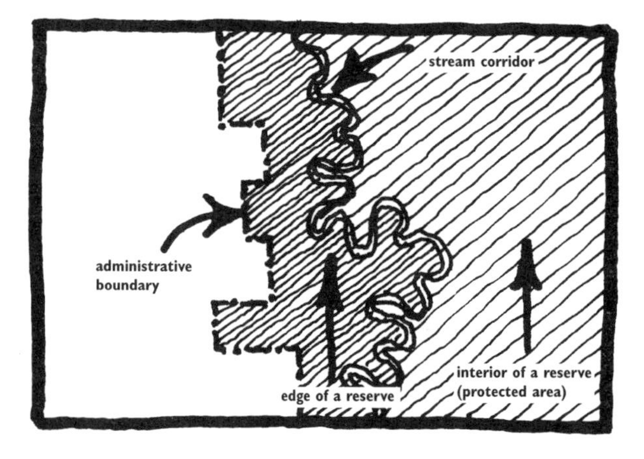

E3. Administrative and natural ecological boundary

Where the administrative or political boundary of a protected area does not coincide with a natural ecological boundary, the area between the boundaries often becomes distinctive, and may act as a buffer zone, reducing the influence of the surroundings on the interior of the protected area.

E4. Edge as filter

Patch edges normally function as filters, which dampen influences of the surroundings on the patch interior.

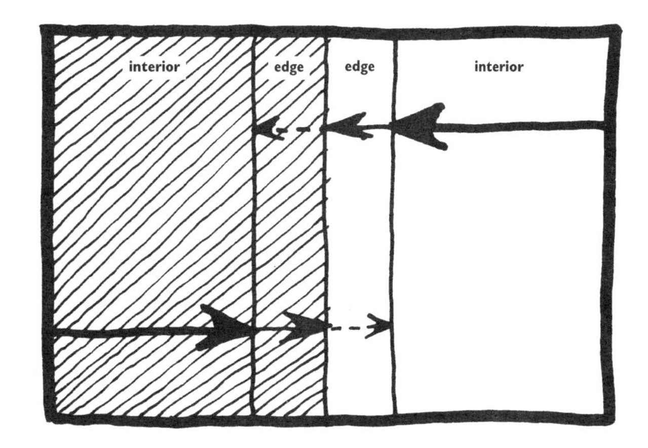

E5. Edge abruptness

Increased edge abruptness tends to increase movement along an edge, whereas less edge abruptness favors movement across an edge.

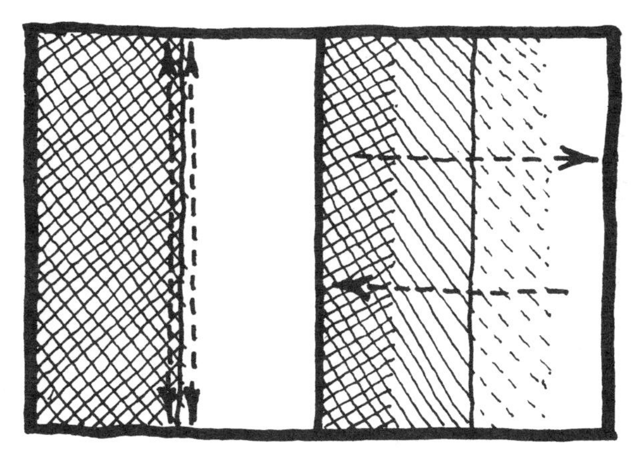

BOUNDARIES: STRAIGHT OR CONVOLUTED?

E6. Natural and human edges

Most natural edges are curvilinear, complex, and soft, whereas humans tend to make straight, simple, and hard edges.

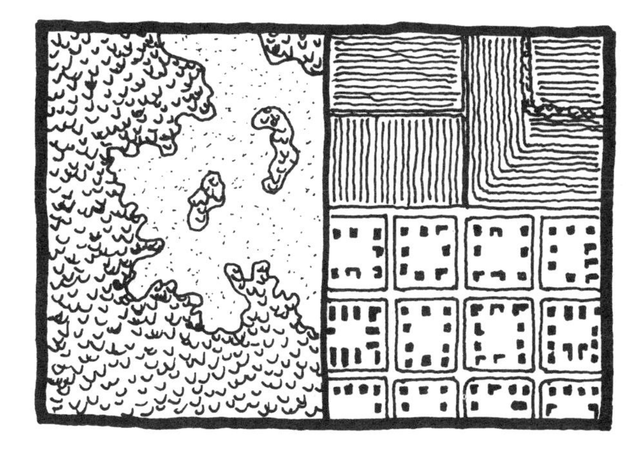

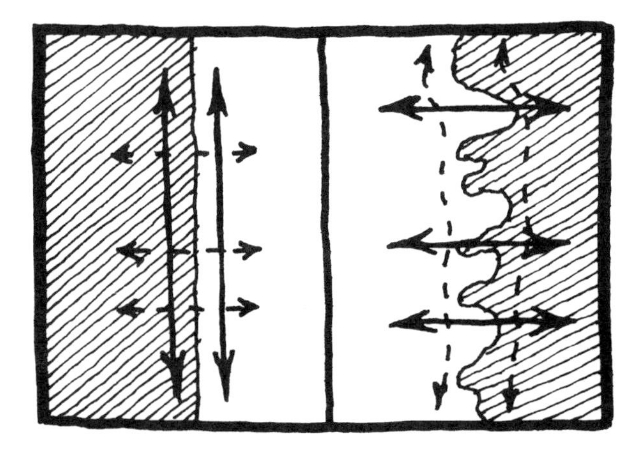

E7. Straight and curvilinear boundaries

A straight boundary tends to have more species movement along it, whereas a convoluted boundary is more likely to have movement across it.

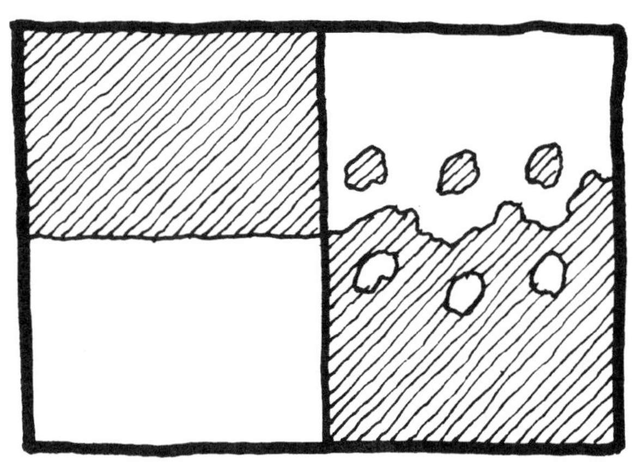

E8. Hard and soft boundaries

Compared with a straight boundary between two areas, a curvilinear "tiny-patch" boundary may provide a number of ecological benefits, including less soil erosion and greater wildlife usage.

E9. Edge curvilinearity and width

Curvilinearity and width of an edge combine to determine the total amount of edge habitat within a landscape.

E10. Coves and lobes

The presence of coves and lobes along an edge provides greater habitat diversity than along a straight edge, thereby encouraging higher species diversity.

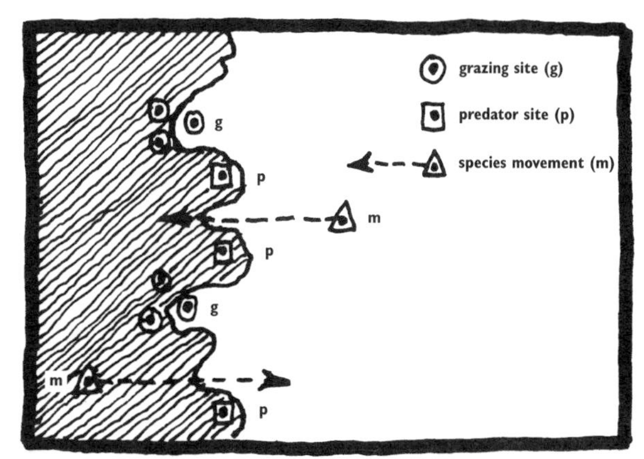

SHAPES OF PATCHES: ROUND OR CONVOLUTED?

E11. Edge and interior species

A more convoluted patch will have a higher proportion of edge habitat, thereby slightly increasing the number of edge species, but sharply decreasing the number of interior species, including those of conservation importance.

E12. Interaction with surroundings

The more convoluted the shape of a patch, the more interaction, whether positive or negative, there is between the patch and the surrounding matrix.

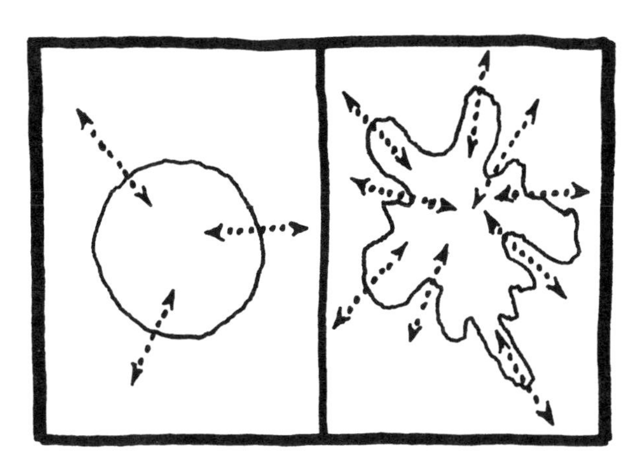

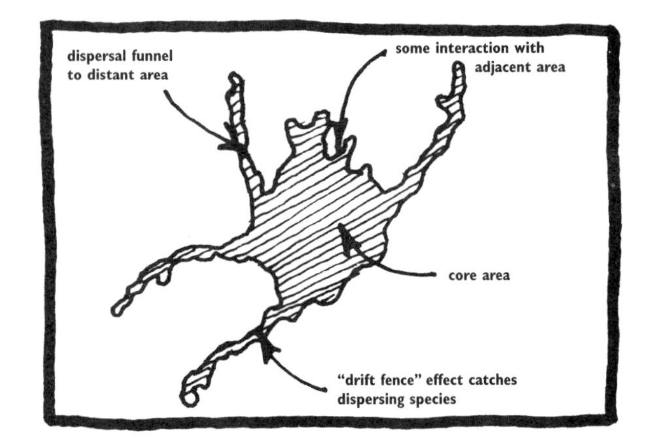

E13. Ecologically "optimum" patch shape

An ecologically optimum patch provides several ecological benefits, and is generally "spaceship shaped," with a rounded core for protection of resources, plus some curvilinear boundaries and a few fingers for species dispersal.

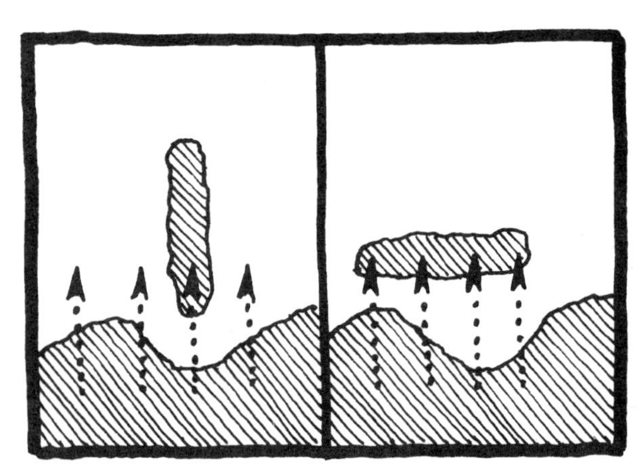

E14. Shape and orientation

A patch oriented with its long axis parallel to the route of dispersing individuals will have a lower probability of being (re-)colonized, than a patch perpendicular to the route of dispersers.

KEY REFERENCES

Forman, R.T.T. 1995. *Land Mosaics: The Ecology of Landscapes and Regions*. Cambridge University Press, Cambridge. [Edge structure, boundary curvilinearity, and patch shape]

Gutzwiller, K.J. and S.H. Anderson. 1992. "Interception of moving organisms: Influences of patch shape, size, and orientation on community structure." *Landscape Ecology* 6, pp. 293-303. [Patch shape]

Hardt, R.A. and R.T.T. Forman. 1989. "Boundary form effects on woody colonization of reclaimed surface mines." *Ecology* 70, pp. 1252-1260. [Boundary curvilinearity]

Harris, L.D. and P. Kangas. 1979. "Designing future landscapes from principles of form and function." In *Our National Landscape: Techniques for Analysis and Management of the Visual Resource*. General Technical Report PSW-34, U.S. Forest Service, Washington, D.C., pp. 725-729. [Patch shape]

Marcot, B.G. and V.J. Meretsky. 1983. "Shaping stands to enhance habitat diversity." *Journal of Forestry* 81, pp. 527-528. [Patch shape]

Milne, B.T. 1991. "The utility of fractal geometry in landscape design." *Landscape and Urban Planning* 21, pp. 81-90. [Boundary curvilinearity]

Morgan, K.A. and J.E. Gates. 1982. "Bird population patterns in forest edge and strip vegetation at Remington Farms, Maryland." *Journal of Wildlife Management* 46, pp. 933-944. [Edge structure]

Ranney, J.W., M.C. Bruner, and J.B. Levenson. 1981. "The importance of edge in the structure and dynamics of forest islands." In Burgess, R.L., and D.M. Sharpe, eds. *Forest Island Dynamics in Man-dominated Landscapes*. Springer-Verlag, New York, pp. 67-96. [Edge structure]

Schonewald-Cox, C. and J.W. Bayless. 1986. "The boundary model: a geographic analysis of design and conservation of nature reserves." *Biological Conservation* 38, pp. 305-322. [Edge structure]

Yahner, R.H. 1988. "Changes in wildlife communities near edges." *Conservation Biology* 2, pp. 333-339. [Edge structure]

See additional references on page 73

The loss and isolation of habitat is a seemingly unstoppable process occurring throughout the modern world. Landscape planners and ecologists must contend with this continuing process if further reductions in biodiversity are to be slowed or halted.

Road corridor including narrow roadsides, Wyoming, U.S.A., R. Forman photo.

Several dynamic processes cause this isolation and loss over time. The key spatial processes include: *fragmentation* (i.e., breaking up a larger/intact habitat into smaller dispersed patches); *dissection* (i.e., splitting an intact habitat into two patches separated by a corridor); *perforation* (i.e., creating "holes" within an essentially intact habitat); *shrinkage* (i.e., the decrease in size of one or more habitats); and *attrition* (i.e., the disappearance of one or more habitat patches).

In the face of continued habitat loss and isolation, many landscape ecologists stress the need for providing landscape connectivity, particularly in the forms of wildlife movement corridors and stepping stones. Despite residual discussion over the effectiveness of corridors in enhancing biodiversity, a growing empirical body of research underlines the positive net benefits accruing from incorporating higher quality linkages between habitat patches.

Powerline corridor, Mississippi, U.S.A., USDA Soil Conservation Service photo.

Corridors in the landscape may also act as barriers or filters to species movement. Some may be population "sinks" (i.e., locations where individuals of a species tend to decrease in number). For example, roadways, railroads, powerlines, canals, and trails, may be thought of as "troughs" or barriers.

Finally, stream or river systems are corridors of exceptional significance in a landscape. Maintaining their ecological integrity in the face of intense human use is both a challenge and an opportunity to landscape designers and land-use planners.

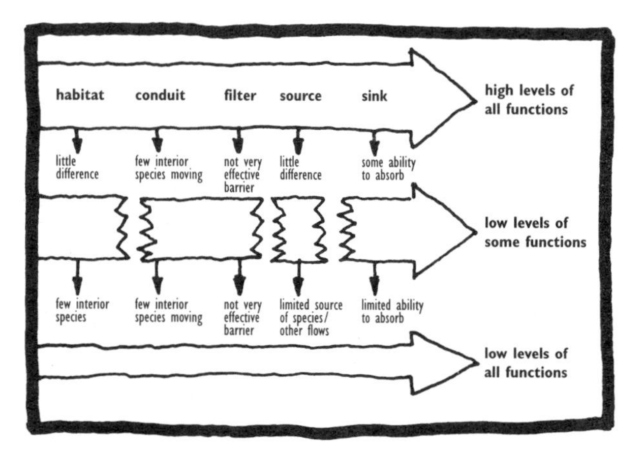

C1. Controls on corridor functions

Width and connectivity are the primary controls on the five major functions of corridors, i.e., habitat, conduit, filter, source, and sink.

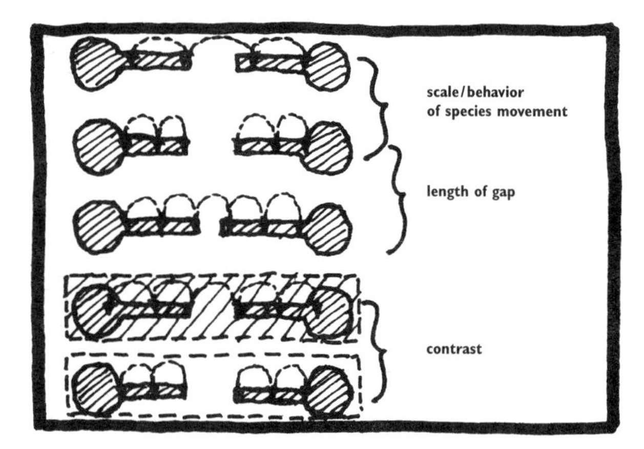

C2. Corridor gap effectiveness

The effect of a gap in a corridor on movement of a species depends on length of the gap relative to the scale of species movement, and contrast between the corridor and the gap.

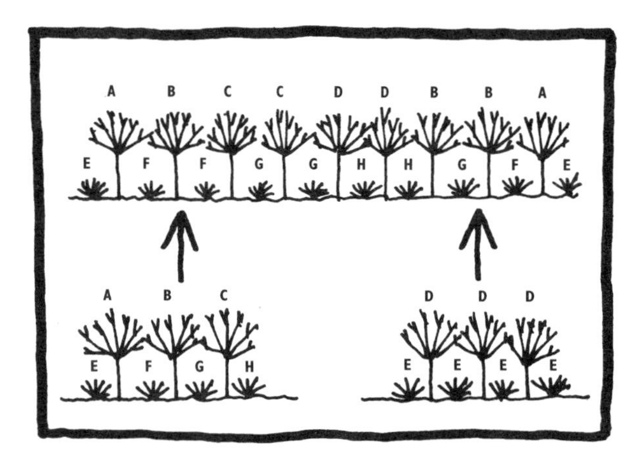

C3. Structural versus floristic similarity

Similarity in vegetation structure and floristics (plant species) between corridors and large patches is preferable, though similarity in structure alone is probably adequate in most cases for interior species movement between large patches.

C4. Stepping stone connectivity

A row of stepping stones (small patches) is intermediate in connectivity between a corridor and no corridor, and hence intermediate in providing for movement of interior species between patches.

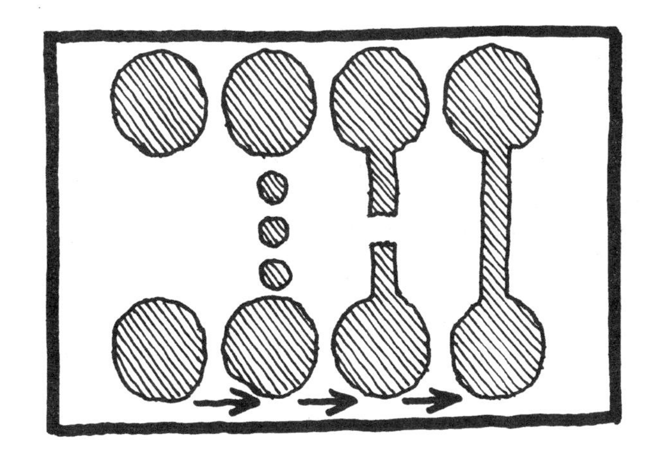

C5. Distance between stepping stones

For highly visually-oriented species, the effective distance for movement between stepping stones is determined by the ability to see each successive stepping stone.

C6. Loss of a stepping stone

Loss of one small patch, which functions as a stepping stone for movement between other patches, normally inhibits movement and thereby increases patch isolation.

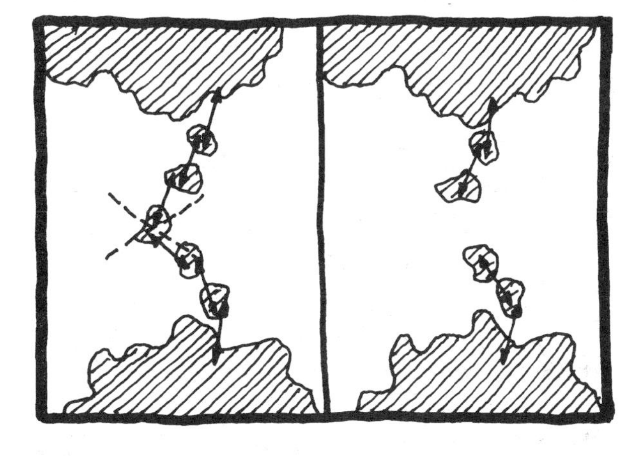

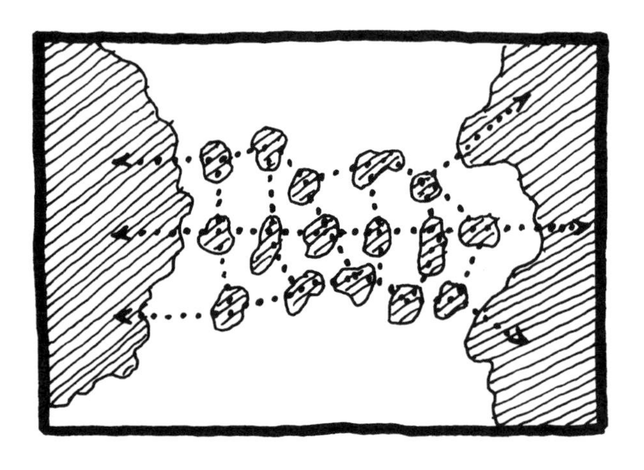

C7. Cluster of stepping stones

The optimal spatial arrangement of a cluster of stepping stones between large patches provides alternate or redundant routes, while maintaining an overall linearly-oriented array between the large patches.

ROAD AND WINDBREAK BARRIERS

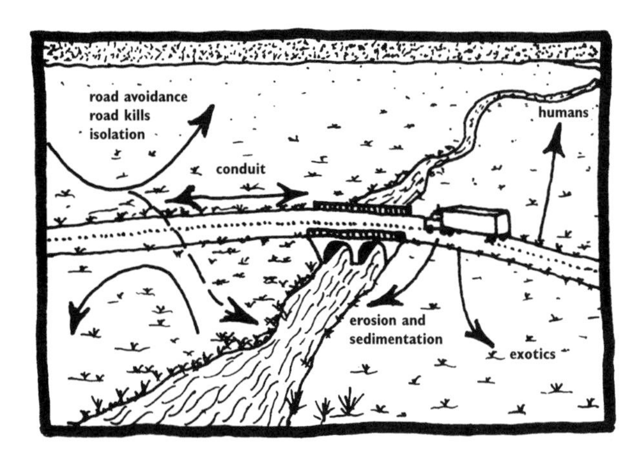

C8. Roads and other "trough" corridors

Road, railroad, powerline, and trail corridors tend to be completely connected, relatively straight, and subject to regular human disturbance. Therefore, they commonly serve as barriers that subdivide populations of species into metapopulations; conduits mainly for disturbance-tolerant species; and sources of erosion, sedimentation, exotic species, and human effects on the matrix.

C9. Wind erosion and its control

Modest winds reduce soil fertility by selectively removing and blowing fine particles long distances, whereas heavier winds often move mid-sized particles only tens of meters. Wind erosion control reduces field size in the preponderant wind direction, and maintains vegetation, furrows, or soil clods, especially in spots susceptible to vortices, turbulence, or accelerated streamline airflow.

C10. Stream corridor and dissolved substances

Dissolved substances, such as nitrogen, phosphorus, and toxins, entering a vegetated stream corridor are primarily controlled from entering the channel and reducing water quality by friction, root absorption, clay, and soil organic matter; these in turn are most effectively provided by a wide corridor of dense natural vegetation.

(1) Contact with plant stems and litter slows water movement

(2) Plant roots absorb dissolved substances prior to reaching the stream

(3) Clay particles hold dissolved substances

(4) Soil organic matter absorbs dissolved substances

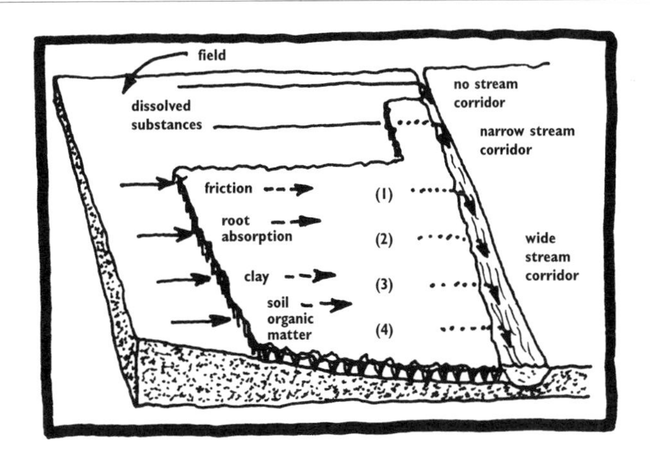

C11. Corridor width for main stream

To maintain natural processes, a 2nd- to ca. 4th-order stream corridor: maintains an interior upland habitat on both sides, which is wide enough to control dissolved-substance inputs from the matrix; provides a conduit for upland interior species; and offers suitable habitat for floodplain species displaced by beaver flooding or lateral channel migration.

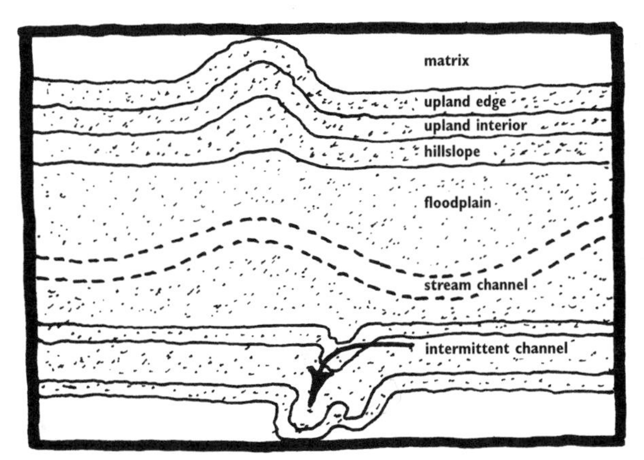

C12. Corridor width for a river

To maintain natural processes, a ca. 5th- to 10th-order river corridor maintains an upland interior on both sides, as a conduit for upland interior species and species displaced by lateral channel migration. In addition, maintaining at least a "ladder-pattern" of large patches crossing the floodplain provides a hydrologic sponge, traps sediment during floods, and provides soil organic matter for the aquatic food chain, logs for fish habitat, and habitats for rare floodplain species.

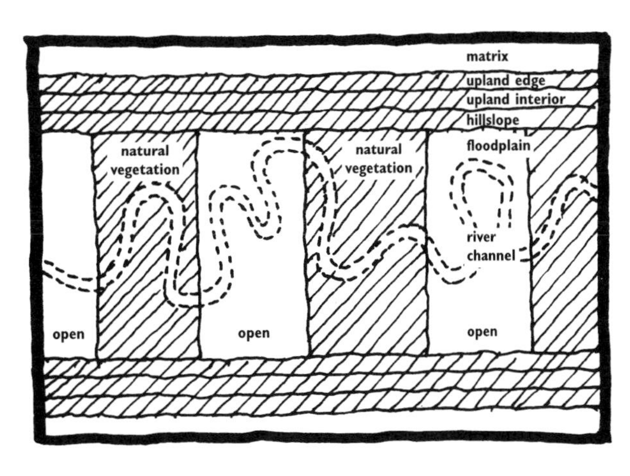

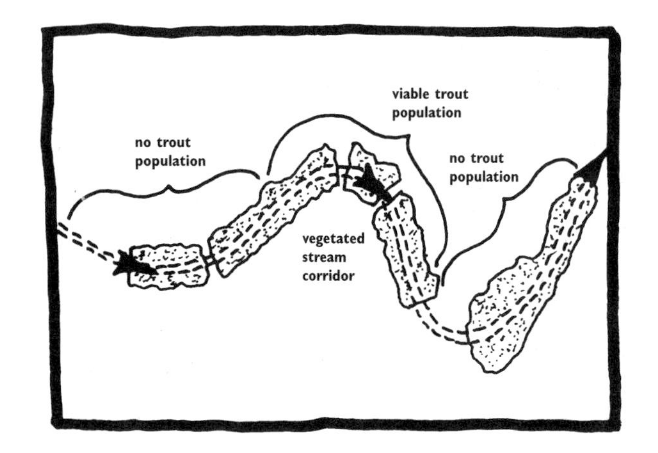

C13. Connectivity of a stream corridor
Width and length of a vegetated stream corridor interact or combine to determine stream processes. However, a continuous stream corridor, without major gaps, is essential to maintain aquatic conditions such as cool water temperature and high oxygen content. Without these, plus other physiological conditions, viable populations of certain fish species, such as trout, will not be maintained.

KEY REFERENCES

Bennett, A.F. 1991. "Roads, roadsides and wildlife conservation: a review." In Saunders, D.A. and R.J. Hobbs, eds. *Nature Conservation 2: The Role of Corridors*. Surrey Beatty, Chipping Norton, Australia, pp. 99-108. [Road barriers]

Binford, M. and M.J. Buchenau. 1993. "Riparian greenways and water resources." In Smith, D. S. and P.C. Hellmund, eds. *Ecology of Greenways. Design and Function of Linear Conservation Areas*. University of Minnesota Press, Minneapolis, Minnesota, pp. 69-104. [Stream and river corridors]

Brandle, J.R., D.L. Hintz and J.W. Sturrock, eds. 1988. *Windbreak Technology*. Elsevier, Amsterdam. (Reprinted from *Agriculture, Ecosystems and Environment* 22-23, 1988). [Windbreaks]

Chasko, G.G. and J.E. Gates. 1982. "Avian habitat suitability along a transmission-line corridor in an oak-hickory forest region." *Wildlife Monographs* 82, pp. 1-41. [Powerline corridors]

Date, E.M., H.A. Ford and H.F. Recher. 1991. "Frugivorous pigeons, stepping stones and weeds in northern New South Wales." In Saunders, D.A. and R.J. Hobbs, eds. *Nature Conservation 2: The Role of Corridors*. Surrey Beatty, Chipping Norton, Australia, pp. 241-245. [Stepping stones]

Forman, R.T.T. 1995. *Land Mosaics: The Ecology of Landscapes and Regions*. Cambridge University Press, Cambridge. [Stream corridor, road and windbreak barriers, and corridor and stepping stones for species movement]

Harris, L.D. and J. Scheck. 1991. "From implications to applications: the dispersal corridor principle applied to the conservation of biological diversity." In Saunders, D.A. and R.J. Hobbs, eds. *Nature Conservation 2: The Role of Corridors*. Surrey Beatty, Chipping Norton, Australia, pp. 189-220. [Corridor for species movement]

Oxley, D.J., M.B. Fenton and G.R. Carmody. 1974. "The effects of roads on populations of small mammals." *Journal of Applied Ecology* 11, pp. 51-59. [Road barriers]

Saunders, D.A. 1990. "Problems of survival in an extensively cultivated landscape: the case of Carnaby's cockatoo, *Calyptorhynchus funereus latirostris*." *Biological Conservation* 54, pp. 277-290. [Stepping stones]

Saunders, D.A. and R.J. Hobbs, eds. 1991. *Nature Conservation 2: The Role of Corridors*. Surrey Beatty, Chipping Norton, Australia. [Corridor for species movement]

See additional references on page 74

cultural functions of a corridor -

habitat - ~~of~~ ~~a few~~ from little difference to few interior species
range -

Species = ~~[illegible]~~ cultural values expressed in a more or less diversified way.

conduit - ~~few~~ species moving ~~or~~ replicated cultural features along the line -

filter - effectiveness of filtering cultural change -

source - ?

sink - ability to absorb culture phenom. species decrease -

livinghomes.

Corridors + Connectivity.

loss + isolation of habitat
fragmentation (breaking up hab. into dispersed patches
dissection (~~breaking up~~ splitting habitat into patches)
perforation (holes)
shrinkage (decrease in size)
attrition (disappearance of patches)

the need for connectivity - biodiversity,
corridors as barriers to species movement.

streams or river systems are c. of exceptional
significance -

---

width + connectivity are the primary controls on the
5 functions of corridors: habitat, conduit, filter
source, + sink.

Corridor gap - movement depends on length of gap.
relative to the scale of species movement + the
contrast between the corridor + the gap.
structural vs floristic similarity - sim. in veg
structure + plant species between corridors +
large patches is preferable.

stepping stones: row of small patches - intermediate
connectivity. Distance between stones a factor.
loss of a stepping stone - inhibits movement, increases isolation
trough corridors: road, railroad, powerline, trail
commonly barriers that subdivide populations -
conduits for disturbance tolerant species; source of
erosion, sedimentation, exotic species, human effects
wind erosion -
stream corridor - dissolved substances, function,
root absorption, clay, organic matter,

corridor width for main stream -

maintain interior or upland habitat both sides -

conduit for upland species, habitat for floodplain species,

river corridor - upland interior, species displaced by lateral channel migration -

ladder pattern = hydrologic sponge, traps sediment, provides organic matter + habitat for rare floodplain species -

connectivity of a stream corridor - essential for aquatic conditions (eg cooler water)

MOSAICS

The overall structural and functional integrity of a landscape can be understood and evaluated in terms of both *pattern* and *scale*. One assay of the ecological health of a landscape is the overall *connectivity* of the natural systems present. Corridors often interconnect with one another to form *networks*, enclosing other landscape elements. Networks in turn exhibit connectivity, *circuitry*, and *mesh size*. Networks emphasize the functioning of landscapes and may be used by planners and landscape architects to facilitate or inhibit flows and movements across a land mosaic.

Hedgerow network with attached woods, England, R. Forman photo.

A common landscape *pattern* is fragmentation, which is often associated with the loss and isolation of habitat. Alternatively, fragmentation is considered as one of several land transformation processes, which together may produce a diminution and isolation of habitat. Fragmentation also results from natural disturbances, such as fires and herbivore invasions, but has become an international land policy issue because of the widespread alteration of land mosaics by human activities.

Mosaic pattern at different scales, Wyoming, U.S.A., R. Forman photo.

The spatial *scale* at which fragmentation occurs is important when identifying strategies to cope with continued habitat loss and isolation. For example, fragmented habitat at a fine scale may be perceived as intact habitat at a broad scale. Only by recognizing and addressing landscape changes across different scales (perhaps at least three) can planners and designers maximize protection of biodiversity and natural processes.

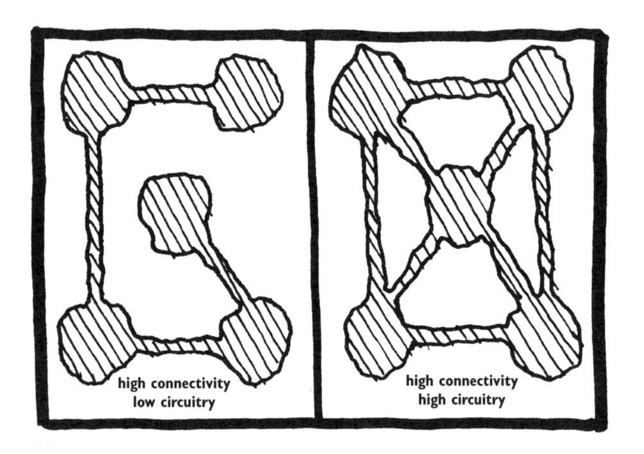

M1. Network connectivity and circuitry

Network connectivity (i.e., the degree to which all nodes are linked by corridors), combined with network circuitry (i.e., the degree to which loops or alternate routes are present), indicates how simple or complex a network is, and provides an overall index of the effectiveness of linkages for species movement.

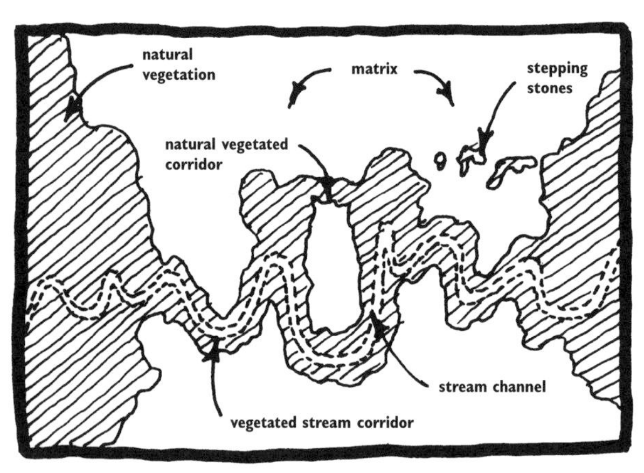

M2. Loops and alternatives

Alternative routes or loops in a network reduce the negative effects of gaps, disturbances, predators, and hunters within corridors, thus increasing efficiency of movement.

M3. Corridor density and mesh size

As mesh size of a network decreases, the probability of survival drops sharply for a species that avoids or is inhibited by the corridors.

M4. Intersection effect

At the intersection of natural-vegetation corridors, commonly a few interior species are present, and species richness is higher than elsewhere in a network.

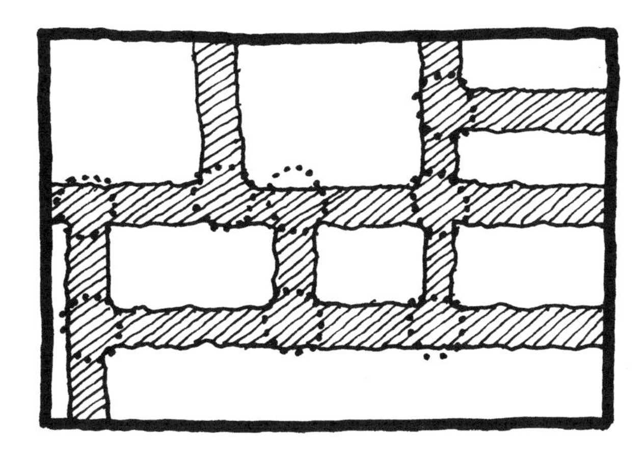

M5. Species in a small connected patch

A small patch or node connected to a network of corridors is likely to have slightly more species and a lower rate of local extinction than an equal-sized patch separated from the network.

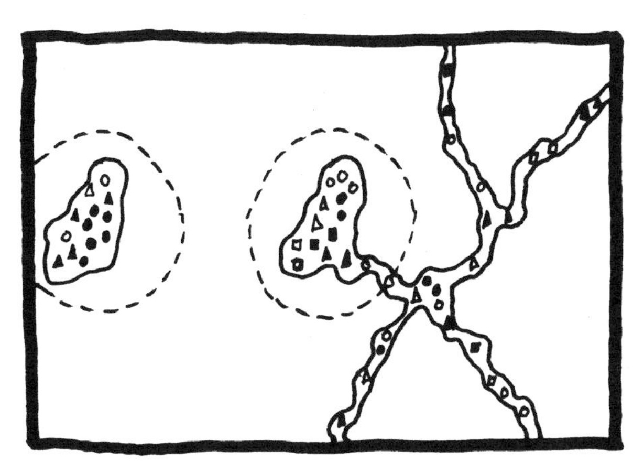

M6. Dispersal and small connected patch

Small patches or nodes along an existing network are effective in providing habitat in which individuals pause and/or breed, resulting in a higher survival rate for dispersing individuals and, hence, more dispersing individuals in the network.

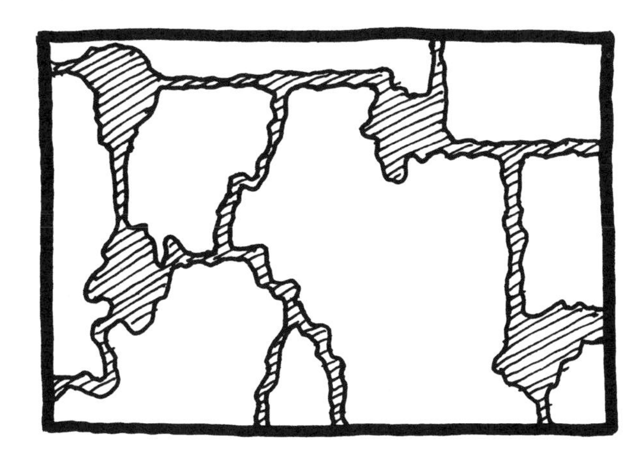

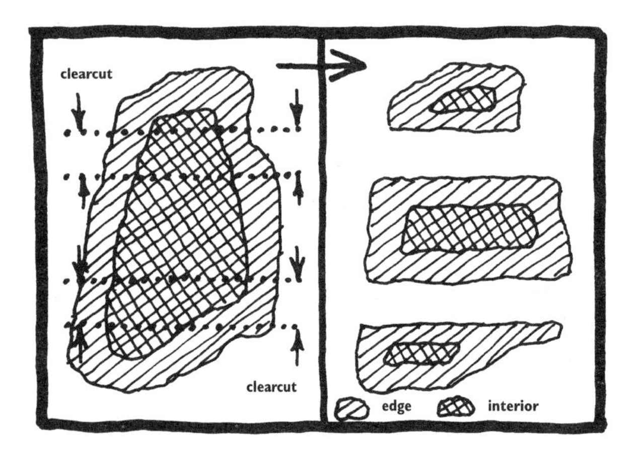

M7. Loss of total versus interior habitat

Fragmentation decreases the total amount of a particular habitat type, but proportionally causes a much greater loss of interior habitat.

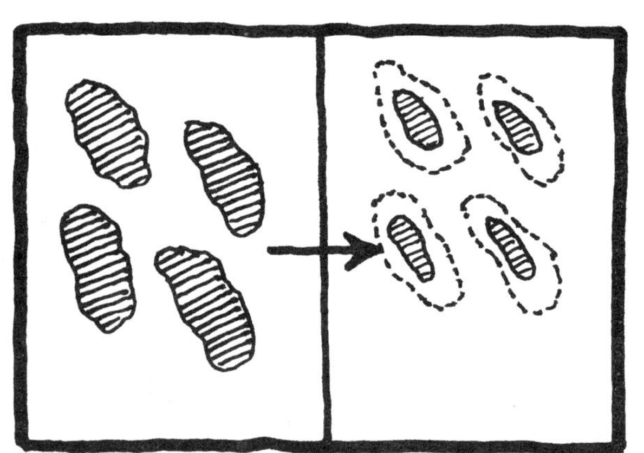

M8. Fractal patches

Fractal configuration is a natural reaction to transition, with isolated patches often reacting similarly to a disturbance as a group. As these patches either become smaller or larger, their structural relationships or pattern stay essentially the same, until an unusually strong disturbance occurs.

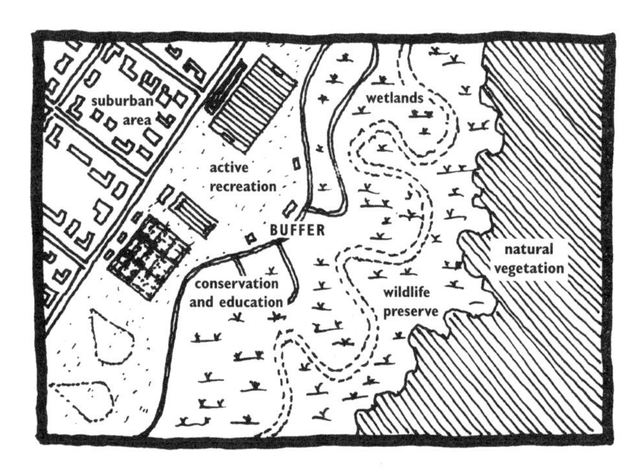

M9. Suburbanization, exotics, and protected areas

In landscapes undergoing suburbanization and consequent invasion of exotic species, a biodiversity or nature reserve may be protected against damage by invaders using a (buffer) zone with strict controls on exotic species.

SCALE: FINE OR COARSE?

M10. Grain size of mosaics

A coarse-grained landscape containing fine-grained areas is optimum to provide for large-patch ecological benefits, multihabitat species including humans, and a breadth of environmental resources and conditions.

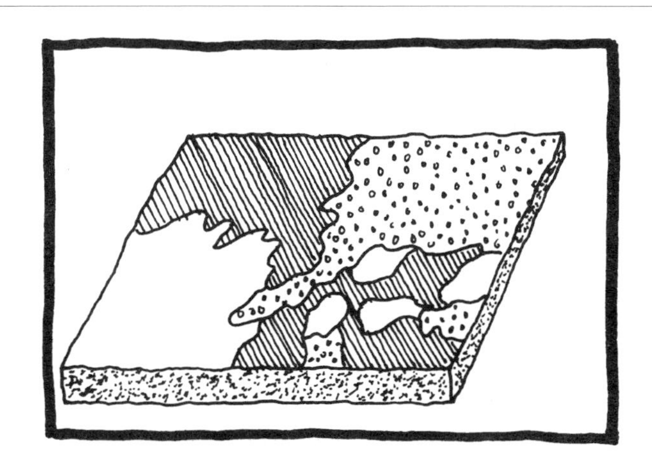

M11. Animal perception of scale of fragmentation

A finely-fragmented habitat is normally perceived as continuous habitat by a wide-ranging species, whereas a coarsely fragmented habitat is discontinuous to all species, except the most wide-ranging large animals.

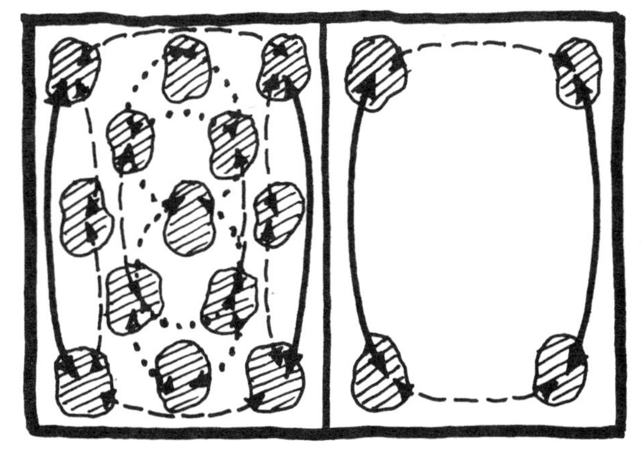

M12. Specialists and generalists

Specialist species are more likely to be negatively affected by fine-scale fragmentation than are generalist species of similar size.

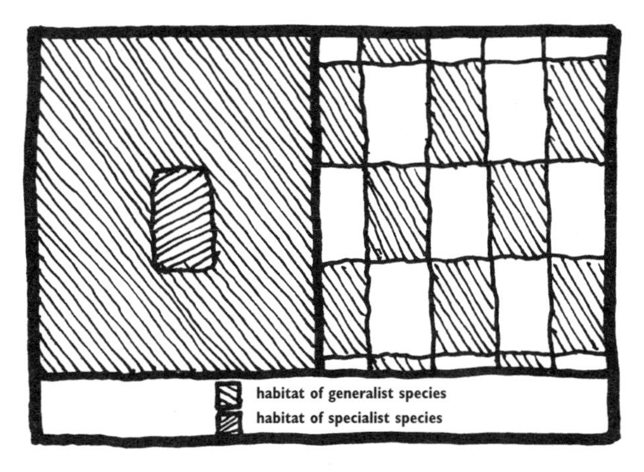

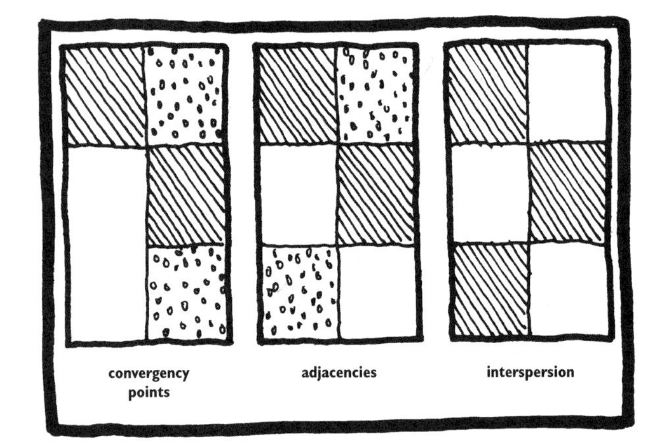

M13. Mosaic patterns for multihabitat species

Multihabitat species are favored by convergency points (junctions where three or more habitats converge), adjacencies (different combinations of adjoining habitat types), and habitat interspersion (habitats scattered rather than aggregated).

KEY REFERENCES

Arnold, G.W. 1983. "The influence of ditch and hedgerow structure, length of hedgerows, and area of woodland and garden on bird numbers in farmland." *Journal of Applied Ecology* 20, pp. 731-750. [Scale]

Forman, R.T.T. 1995. *Land Mosaics: The Ecology of Landscapes and Regions.* Cambridge University Press, Cambridge. [Networks, fragmentation, and pattern]

Forman, R.T.T. and J. Baudry. 1984. "Hedgerows and hedgerow networks in landscape ecology." *Environmental Management* 8, pp. 495-510. [Networks]

Franklin, J.F. and R.T.T. Forman. 1987. "Creating landscape patterns by cutting: ecological consequences and principles." *Landscape Ecology* 1, pp. 5-18. [Fragmentation and pattern]

Hobbs, R.J. 1995. Landscape ecology. *Encyclopedia of Environmental Biology* 2, pp. 417-428. [Fragmentation and pattern]

Knaapen, J.P., M. Scheffer and B. Harms. 1992. "Estimating habitat isolation in landscape planning." *Landscape and Urban Planning* 23, pp. 1-16. [Fragmentation and pattern]

Landers, J.L., R.J. Hamilton, A.S. Johnson and R.L. Marchington, 1979. "Foods and habitat of black bears in southeastern North Carolina." *Journal of Wildlife Management* 43, pp. 143-153. [Fragmentation and pattern]

Lyon, L.J. 1983. "Road density models describing habitat effectiveness for elk." *Journal of Forestry* 81, pp. 592-595. [Networks]

Noss, R.F. and L.D. Harris. 1986. "Nodes, networks, and MUMs: preserving diversity at all scales." *Environmental Management* 10, pp. 299-309. [Networks and scale]

Saunders, D.A. 1989. "Changes in the avifauna of a region, district and remnant as a result of fragmentation of native vegetation: The wheatbelt of Western Australia. A case study." *Biological Conservation* 50, pp. 99-135. [Fragmentation and scale]

See additional references on page 77

OVERVIEW

A myriad of complex and seemingly unrelated decisions occurs in the land-use planning and landscape architecture professions. During the analysis phase of a project, a variety of social, legal, demographic, topographic, microclimatic, and other site-specific information is simultaneously considered. Too seldom, however, does site analysis incorporate a more broad-reaching, landscape ecological approach, where the impacts of a particular land-use plan or landscape design are considered within the larger, ecological context of the landscape or region.

The following section presents a series of (1) hypothetical or schematic applications, and (2) actual case studies (in brief). Together these illustrate how land-use planners and landscape architects may (or have) incorporate(d) landscape ecological principles in their work. A range of scales and types of projects is illustrated.

Schematic or conceptual landscape changes are applied in six cases over a broad range of spatial scales. These applications state a problem or proposed change in the landscape, and portray "better" and "worse" designs, together with rationales.

A prototypical landscape type with a mix of agricultural, suburban, and forested areas is used in the first section. A variety of landscape ecological elements, such as habitat patches, stream corridors, wildlife movement corridors, roads and powerlines, natural edges and boundaries and artificial edges, is illustrated. This heterogeneous landscape type, plus the method of representation, is widespread, for example in many parts of the U.S., Europe, South America, and Russia. In addition, human-induced changes and developments are frequently planned and occurring in most areas around the world.

The principles applied to this typical agricultural-suburban-forested area are just as valid in a coastal, desert, or mountainous area. Land-use plans and landscape designs incorporate changes of a generalizable nature. What matters more than the specific land-use change or design proposal are the consequences of that change or design. For example, does the design proposal create a gap in an animal movement corridor? Does the land-use plan reduce the size or change the shape of an important patch of natural habitat? Furthermore, the same set of landscape ecological principles applies, whether working at the scale of a residential site or a regional park.

Specific case studies have been selected to illustrate how landscape ecological principles have been applied within actual or proposed projects. These case studies are taken from scientific, planning, and other journals. The projects are from around the world and have in some way incorporated landscape ecology in their conception and implementation. In some cases empirical evidence indicates how the incorporation of these principles into a design or plan has either improved or maintained the integrity of the ecological functioning of a particular landscape or region.

It is noteworthy that land-use planners and landscape architects are indeed empowered to make meaningful contributions to the overall ecological health of their environment, by incorporating a landscape ecological approach in their work. Over time, numerous examples of such landscape-ecologically based work will no doubt be documented. Hopefully, readers of this book will be among those whose works are so publicized.

SCHEMATIC APPLICATIONS

The following six schematic applications are selected to illustrate landscape ecological principles over a range of scales, from a macro or regional scale, to a micro or site scale. One of the powerful messages of illustrating the applications across a range of scales is that these principles are all applicable and effective independent of the size of the project. The six examples are:

Macro or regional scale

A regional wildlife conservation park

A new suburban development project

Meso or landscape scale

A new road

An urban park

Micro or site scale

A cluster of backyard gardens

A wildlife movement corridor

MAP AND KEY OF EXISTING CONDITIONS

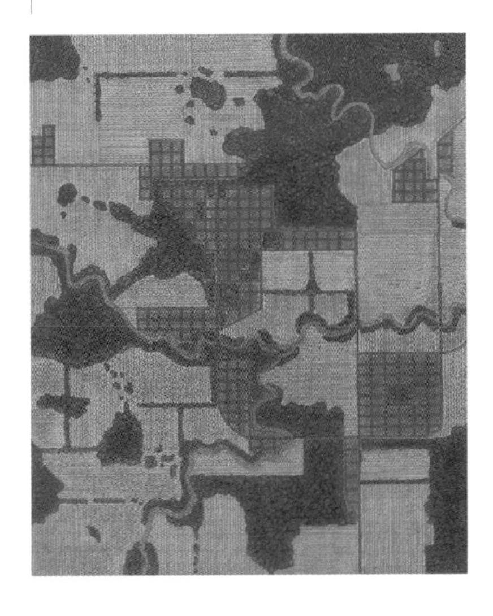

not to scale

Agricultural Fields

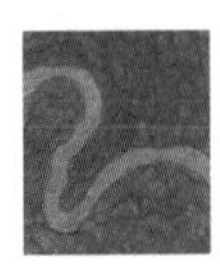

River/Stream

Forested/Wooded

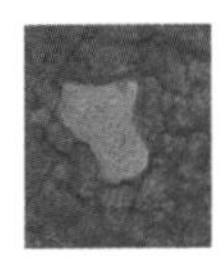

Water Body

Subdivision/ Development/ Street

The problem:
Overall, where to locate a regional park and, specifically, how to allocate land-uses of different intensity within the park

The regional area

Mixed urban/suburban development and natural forested areas within agricultural matrix

The study area

Largest patch of natural forest close to largest area of suburban development

A "better" design

1. Conservation area (1) mostly circular in shape
2. Passive-use area (2) as buffer between conservation and active-use area (3)
3. Active-use area (3) close to suburbanization

A "worse" design

1. Conservation area (1) less circular in shape
2. Passive-use area (2) does not provide a continuous buffer
3. Entirety of lake in active-use area (3)
4. Active-use area closer to conservation area

MACRO OR REGIONAL SCALE

A new suburban development

The problem:
Where to ideally locate a specified amount of suburban development

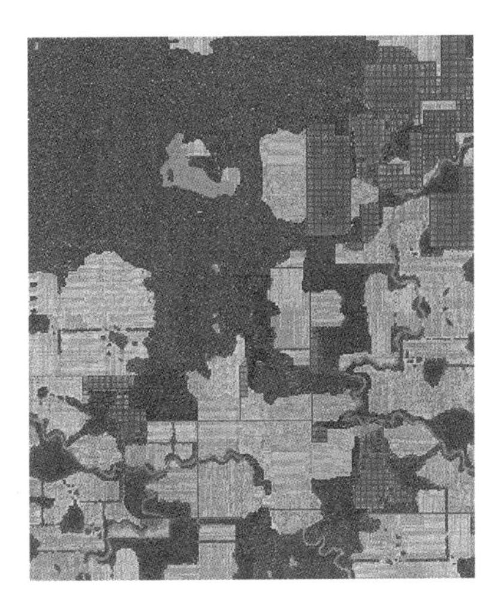

The regional area

Mixed urban/suburban development and natural forested areas within agricultural matrix

The study area

Two options considered:

1) within large forested patch; and

2) close to existing suburban development

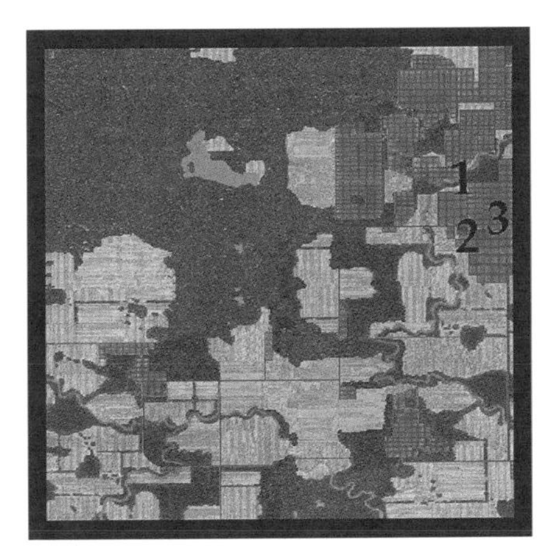

A "better" design

1. River corridor made narrower, although not completely "broken"
2. Forested patch size reduced, and patch surrounded by suburbanization
3. New suburban development concentrated within area of existing suburbanization

A "worse" design

1. Species movement around lake obstructed
2. Nutrient enrichment, other pollution in lake
3. Invasion of exotics into larger forested area (previously undisturbed)
4. Forested corridor broken by road and roadway

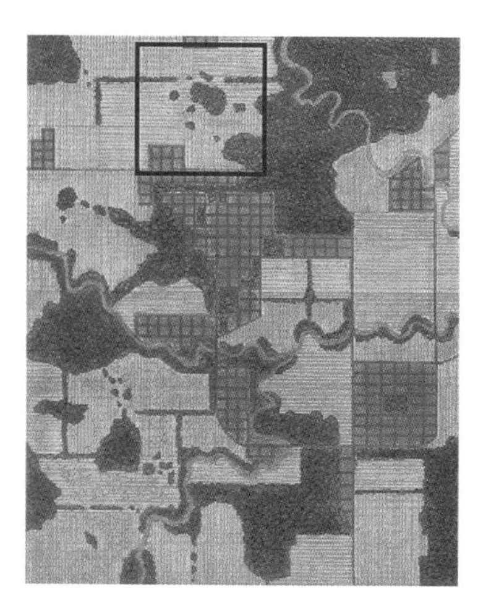

The regional context

Mixed urban/suburban development and natural forested areas within agricultural matrix

The study area

Agricultural fields interspersed with small patches of remnant natural vegetation

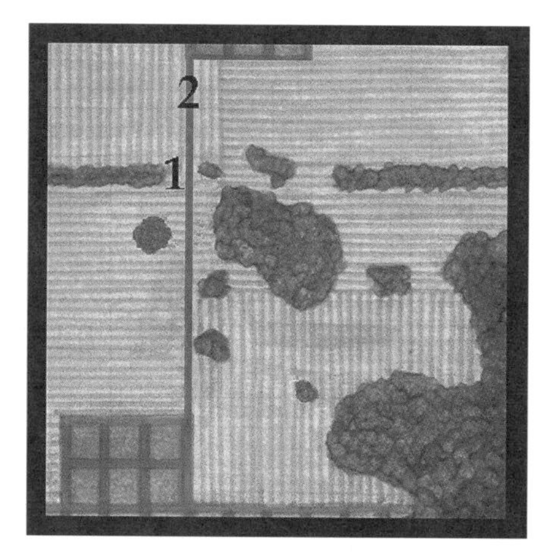

A "better" design

1. Barrier added between hedgerows
2. Roadside exotics spread to fields

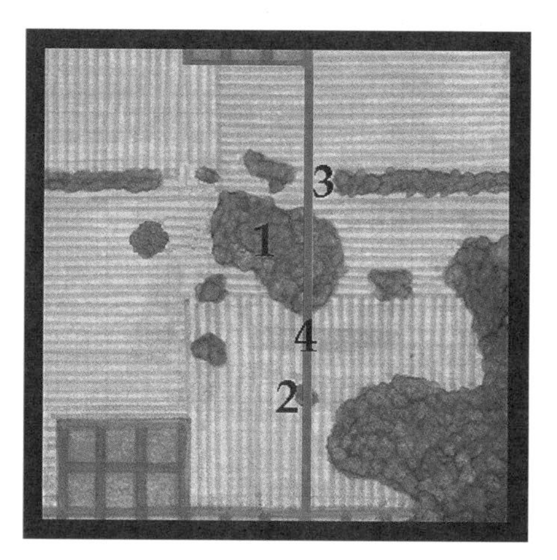

A "worse" design

1. Largest "local" patch bisected
2. Small patch bisected/eliminated
3. Barrier added between hedgerows
4. Roadside exotics spread in woods and fields

MESO OR LANDSCAPE SCALE

Suburban park land expansion

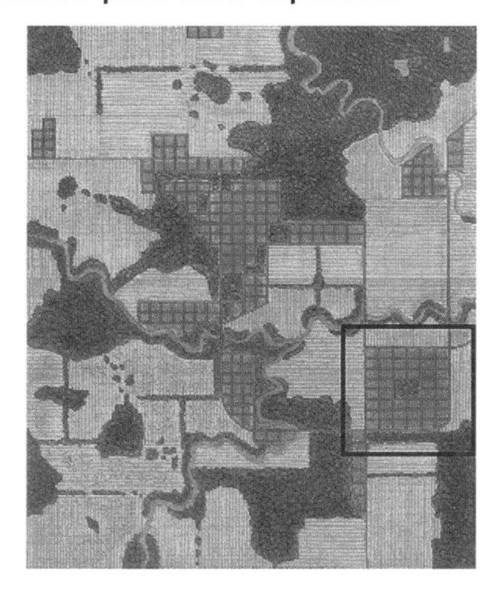

The regional context

Mixed urban/suburban development and natural forested areas within agricultural matrix

The study area

Suburban park land expansion

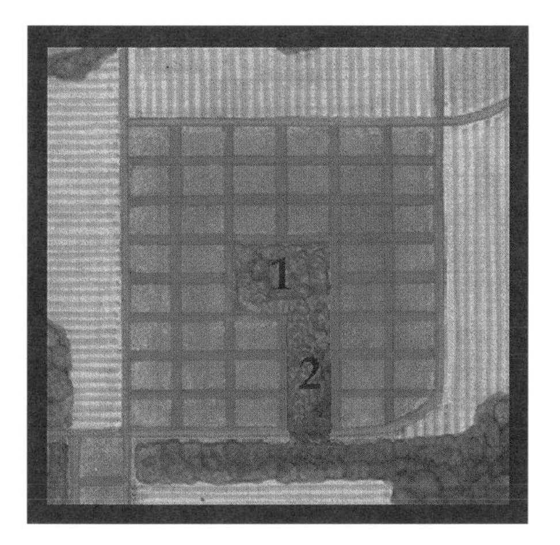

A "better" design

1. Existing park land: a net increase in total area
2. Park linked to existing, natural vegetated corridor at edge of suburban area

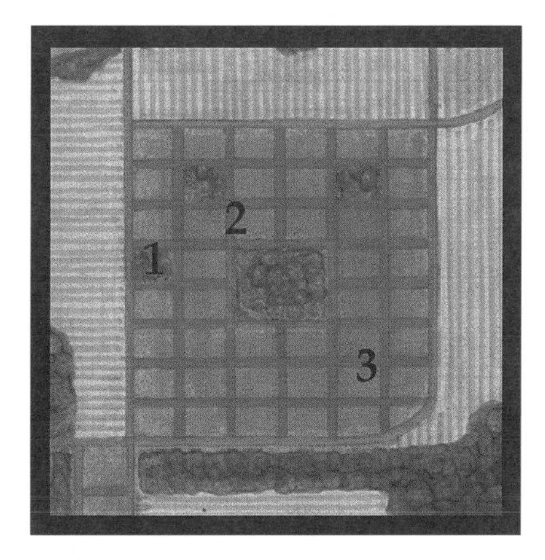

A "worse" design

1. A series of small, "pocket" parks added, mainly near edge of development
2. Barrier between new parks and existing larger park
3. No connections to areas of natural vegetation outside suburbanization

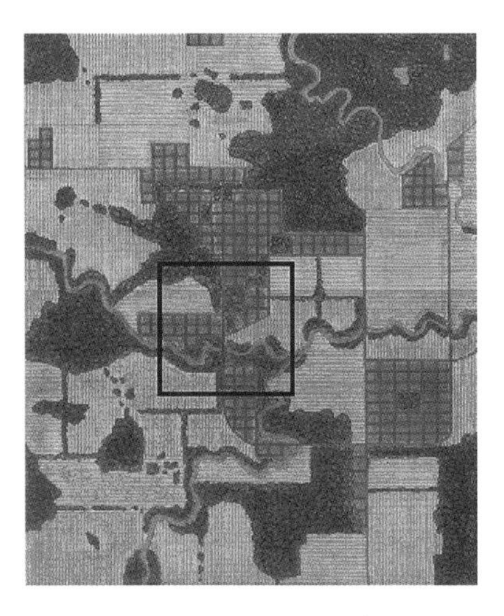

The regional context

Mixed urban/suburban development and natural forested areas within agricultural matrix

The study area

Suburban housing development adjacent to protected riparian corridor/habitat

A "better" design

1. Continuous vegetated corridor maintained through backyards of houses
2. Housing setbacks at minimum distance from road/maximum from corridor
3. Use of native vegetation/minimal threat due to spread of exotic species

A "worse" design

1. Vegetated corridor narrow and broken; less movement of key species
2. Housing setbacks maximum from road/minimum from forested corridor
3. Increased threat of disturbance to natural vegetation due to use/spread of exotics

MICRO OR SITE SCALE

A wildlife movement corridor

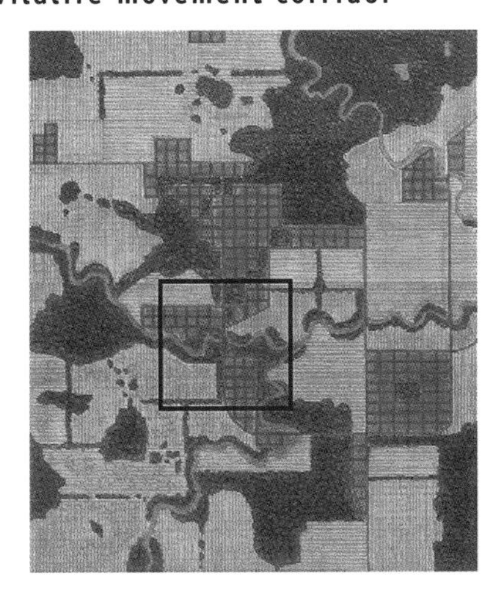

The regional context

Mixed urban/suburban development and natural forested areas within agricultural matrix

The study area

Intersection of roadway, riparian corridor, and agricultural fields

A "better" design

1. Placement of roadway bridge allows for species movement below on both sides of riparian corridor
2. Native vegetation left intact to provide continuous corridor

A "worse" design

1. Placement of roadway bridge does not permit species movement on either side of riparian corridor
2. Native vegetation eliminated, causing gap in movement corridor

CASE STUDIES IN BRIEF

The following case studies similarly represent a range of scales and diverse types of landscapes from around the world. Case studies include both "sound" and "unsound" landscape ecology-based planning. A practical use of landscape ecology in land-use planning and landscape architecture of course does not assure success. Much can be learned from these brief descriptions of projects incorporating landscape ecology principles at different scales, but of course more is available in the accompanying references.

1. MESO SCALE. River corridor: Kissimmee River wetlands, channel, and floodplain restoration (Florida, U.S.A.)

After a period of twenty years, the negative ecological effects of channeling the Kissimmee River have become increasingly clear. The original beneficial objectives of channelization (i.e., flood control, improved navigation, and more land area for cattle grazing) have been outweighed by the costs of the project: (1) change from a meandering river to a deep canal with little biological use; (2) lowering of the surrounding water table, with a drying up of most of the floodplain wetlands; (3) degradation of remaining wetlands due to maintenance of constant water levels; and (4) alteration of the seasonal pattern of water flow critical to fish, resident wildlife, and migratory waterfowl.

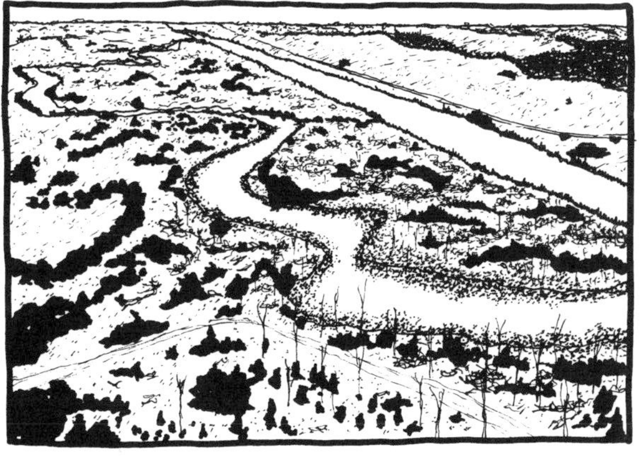
Channelized river and restored floodplain.

This restoration project has just begun and includes a series of recommendations designed to help regain the ecological structure and functioning of a pre-existing riparian mosaic and system in Central Florida. If completed, the project will be the largest restoration of a riparian corridor and associated wetlands and floodplain ever undertaken. The feasibility of restoring most of the critical hydrological functioning of the system has been tested in a demonstration area. The three primary goals of the overall project are to restore: (1) water quality; (2) water level fluctuations; and (3) natural resource values. By focusing on the functioning of the system, i.e., flows of water and movements of animals, it is hoped that the structure or pattern of the landscape will rapidly reform.

Reference:
Karr, James R., 1988. "Kissimmee River: Restoration of Degraded Resources." *Proceedings Kissimmee River Restoration Symposium*, S. Florida Water Mgmt. Dist., W. Palm Beach, FL.

(Also see one issue of *Landscape and Urban Planning*, 1995)

2. MACRO SCALE. Regional corridor: Mountains to Sound Greenway, A Recreational/Nature Preserve (Washington, U.S.A.)

This regional greenbelt proposal is a master planning framework for how development and conservation may coexist within an area of rapidly expanding suburbanization near Seattle. The primary strategy is to preserve and link the remaining native habitat patches, via a greenway approximately 100 miles long. This incorporates both human recreational areas and increasingly wild areas that help protect the biological diversity of the region.

This proposal illustrates several of the landscape ecology principles discussed herein. First, the planning for conservation of multiple recreational reserves provides a series of "buffer" zones between undisturbed, natural habitat and human development. Second, the series of "stepping stone" reserves, interconnected via a continuous network of corridors, both encourages faunal movement and provides protected habitat for a diversity of species.

Reference:
Kobayashi, Koichi, ed., 1995. "Jones & Jones: Ideas Migrate...Places Resonate." *Process Architecture*, no. 126, p. 49.

3. MESO SCALE. Large patches and mosaics: "A New Forestry Application" (Maine, U.S.A.)

This proposal offers a landscape ecologically-based strategy for better managing timber production within the Maine spruce-fir forest. Two underlying principles or techniques are pinpointed: (1) create and maintain vertical diversity in the forest canopy; and (2) ensure that the biological "legacy" of old-growth forest is transferred to regenerating stands.

These underlying principles are translated into a series of specific and strategic practices to maintain a more ecologically-sound mosaic of land uses. The basic goal is to maintain the biological diversity of the region.

At the landscape level three forest patch types are identified: (1) high yield plantations; (2) lower yield (i.e., "new forestry") areas; and (3) reserves. The proposal calls for the distribution of these three areas across the landscape as large blocks, creating a coarse-grained mosaic. The new forestry areas act as buffers between the reserves and the high-yield plantations. Furthermore, a range of harvest sizes are recommended, which are spatially arranged according to the existing configuration and fragmentation, as well as social values and aesthetics.

Reference:
Seymour, Robert S. and Malcolm L. Hunter, Jr., 1992. "New Forestry in Eastern Spruce Forests: Principles and Applications to Maine," *Pub. No. 716*, Orono, ME.

the triad concept of forest land allocation

ecological reserves

new forestry

high yield plantations

arrangement of the triad on the landscape

Forestry and reserves arranged on the land.

4. MACRO SCALE. Corridor: International Wildlife Corridor for Deer (Italy/Switzerland)

An international linkage connecting two wildlife reserves (one in Italy, the other in Switzerland) has been created to maintain and restore the annual migration corridor of red deer (similar to elk).

The combined area of the two reserves exceeds 1,000 square kilometers, with the corridor comprising nearly 150 square kilometers. and traversing a gradient of elevations. The migratory patterns of the red deer involve movement between open meadows in lower elevations during the winter, and higher mountainous elevations during the summer.

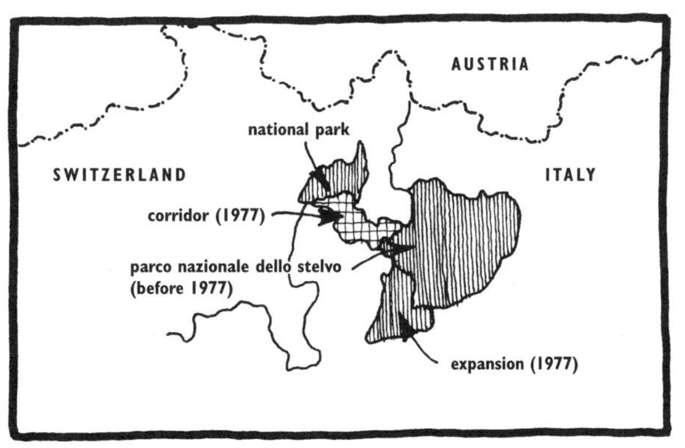

International corridor for deer.

Reference:
Harris, L. D., and J. Scheck, 1991. "From implications to applications: the dispersal corridor principle applied to the conservation of biological diversity." In Saunders, D. A., and R. J. Hobbs, eds., *Nature Conservation 2: The Role of Corridors*. Surrey Beatty, Chipping Norton, pp. 189-220.

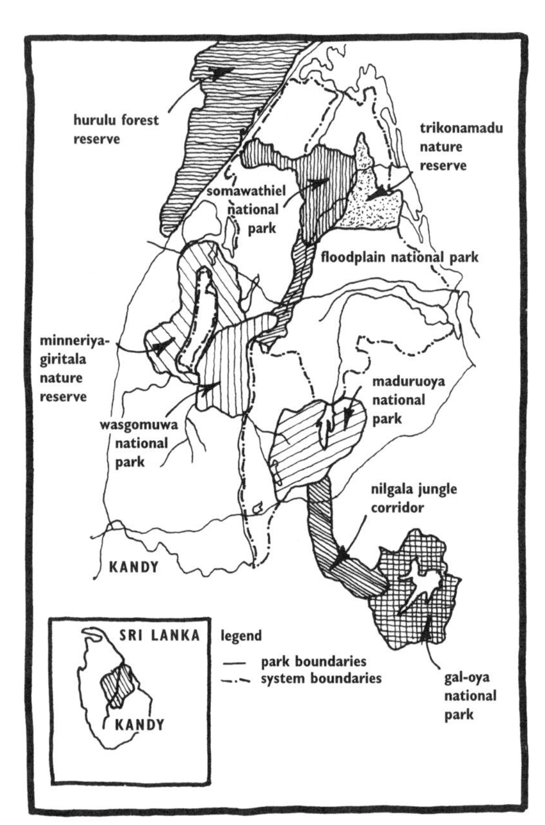

Network for elephants.

5. MESO SCALE. Network: Elephants and other Forest Species (Sri Lanka)

A well-researched and integrated faunal corridor system has been established for elephants and other forest species in Sri Lanka. The map illustrates preserves and the corridor linkages between them. The combined area of the protected network equals nearly 20% of the total land area of the entire country. The series of protected areas provides a diversity of required habitats for reproduction, foraging, and seasonal movement of elephants, similar to the Swiss/Italian red-deer corridor case cited above.

Reference:
Harris, L. D., and J. Scheck, *ibid.*, pp. 189-220.

6. MESO SCALE. Road crossing: Faunal Underpasses along Highway (Southern Florida, U.S.A.)

Nearly 30 underpasses were built for wildlife and waterflows beneath Interstate Highway 75 (Alligator Alley), which links Naples with Ft. Lauderdale and Miami, and crosses the northern edge of the Everglades. This is the largest wildlife underpass program in the world, and additional smaller underpasses are currently under construction on nearby state roads. Construction of the wide highway might have split in two one of the state's largest and most important habitats, without this proper design consideration.

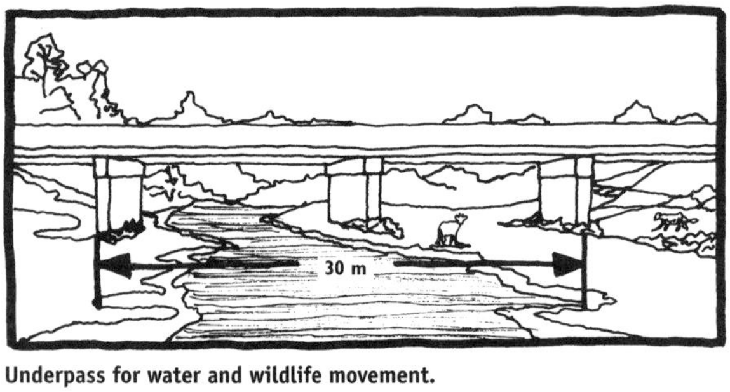

The major objective was to reduce roadkills of the endangered Florida panther. Numerous panther crossings in the underpasses have been recorded, and roadkills have decreased. As yet the effect on the population size of this rare species is unknown. An additional benefit of the underpasses is that many species,

30 m

Underpass for water and wildlife movement.

including deer, alligators, and bobcats, have used them. The underpasses have proven to be an effective mitigation strategy to reduce the barrier and isolating effect of roads.

Reference:
Smith, D. S. 1993. "Greenway case studies." In Smith, D. S. and P.C. Hellmund, eds., *Ecology of Greenways. Design and Function of Linear Conservation Areas.* Univ. of Minn. Press, Minneapolis, pp. 161-208.

7. MICRO SCALE. Wildlife tunnel: movement of an endangered species (New South Wales, Australia)

The habitat of an endangered marsupial species, the mountain pygmy-possum, in southeastern Australia, was fragmented by a major new road. Construction of a subterranean tunnel successfully reconnected the habitat, to accommodate the unique and specific habitat requirements of the species.

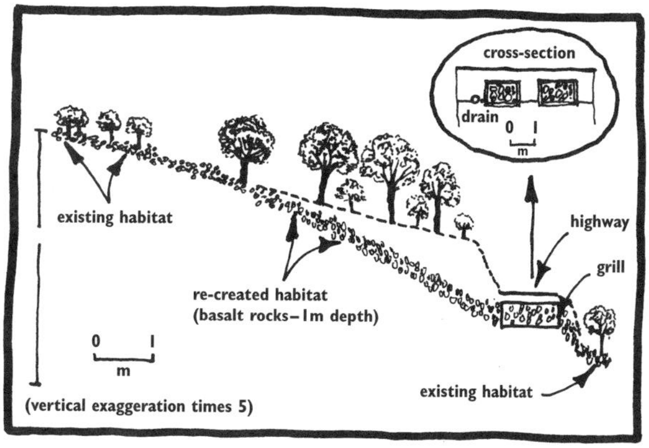

This wildlife movement corridor was constructed to imitate the native habitat of the species. It assisted in the normal seasonal dispersal of the population, which had been disrupted by the road. Following construction of the tunnel, population survival and dispersal rates of the species in this disturbed and disconnected area were similar to those of the species in a nearby undisturbed area.

Wildlife Tunnel constructed beneath road.

As habitats are increasingly bisected and fragmented by roads and development, artificial links such as underpasses, tunnels, and overpasses between fragments must be carefully considered. Knowledge of the habitat requirements and social organization of the key species is critical. One can then determine whether there is any strong reason not to plan and design artificial links as a suitable management strategy.

Reference:
Mansergh, I. M., and D.J. Scotts, 1989. "Habitat continuity and social organization of the mountain pygmy-possum restored by tunnel." *Journal of Wildlife Management* 53; pp. 701-707.

8. MICRO SCALE. Amphibian tunnels permitting seasonal reproductive movement (Germany and elsewhere)

Many amphibians must move from upland habitats to a pond or other water body to accomplish their reproductive cycle. Roads often separate the pond from the upland. The effectiveness of alternative designs varies for facilitating the seasonal movement of amphibians beneath roads in Germany and other European countries, plus in the U.S.A.

Amphibian Tunnel inserted into road surface.

Many tunnel projects have been judged as failures, e.g., because of excessive mortality, predation, inadequate light or ventilation, filling with water, lack of light at end of tunnel, and poorly designed "drift fences" leading amphibians to the tunnel entrance.

However, several designs that address these issues are successful in enabling such movement. Most important is that tunnels provide two-way access between habitat on both sides of the roadway. Many unsuccessful designs allow only one-way crossings. A secondary design consideration is the allowance for light and air into the subterranean tunnels, without which many amphibians are not successful in completing their crossings. Several studies indicate that road closures during peak seasonal, spawning-related amphibian movement are especially successful in reconnecting fragmented habitat.

Reference:

Langton, T. E. S., ed., 1989. "Amphibians and roads," ACO Polymer Products, Bedfordshire, UK.

9. MESO SCALE. Patch location and size: New forest/timber plantations (The Netherlands)

New forested areas are being located for biodiversity, recreation, and timber harvest. The two most important aspects to consider, vis-à-vis preserving and enhancing biological diversity, are: (1) area of a wooded patch and the population dynamics of key area-dependent species; and (2) arrangement of wooded patches in the surrounding matrix, especially the total area of surrounding woodlots and the distance to the nearest woodlot.

Several large forests rather than many small ones are determined to be optimum. Using a computer simulation model, this study pinpoints where new forest patches would best be located to produce population increases of key selected species. Large new patches are to be established in areas of farmland with a reasonable density of scattered woods. These existing patches act as stepping stones for long distance dispersal of species, as well as population sources that help sustain species in the newly created forest patches.

Reference:
Harms, W. B., and P. Opdam, 1989. "Woods as habitat patches for birds: application in landscape planning in the Netherlands." In Zonneveld, I. S., and R.T.T. Forman, eds., *Changing Landscapes: An Ecological Perspective*, Springer-Verlag, N.Y.

10. MESO SCALE. Regional network: Recreation and habitat/flood protection corridors (Southeastern Wisconsin, U.S.A.)

As one of the most extensive greenway networks in the U.S., this seven county district in southeastern Wisconsin has 467 square miles of proposed protected corridors. Four types of linear elements are included: (1) former railroad beds; (2) riparian zones; (3) agricultural riparian zones; and (4) ridgelines. The network extends across a mosaic of different land-use types, from urban to suburban to rural, and includes scenic as well as important natural resources for wildlife conservation purposes.

The major objectives of this extensive system are: (1) habitat protection; (2) recreation; and (3) floodwater protection and control. The procedure for selecting corridors is based on evaluating a variety of ecological patterns and processes. Corridor width is considered most important. These environmental corridors provide increased connectivity for human and wildlife movement and link major nodes, i.e., protected parks within the region. The network forms alternative routes, providing round trips for human recreation and permitting moving wildlife to avoid disturbance. The focus on riparian corridors helps provide sponges for flood protection.

Reference:
Smith, D. S. 1993. "Greenway case studies." In Smith, D. S. and P.C. Hellmund, eds., *Ecology of Greenways. Design and Function of Linear Conservation Areas*. Univ. of Minn. Press, Minneapolis, pp. 161-208.

11. MESO SCALE. Wildlife network: Santa Monica Mountains to Santa Susana Mountains Corridors (California, U.S.A.)

This regional effort, occurring in one of the most rapidly urbanizing areas in the U.S., is to link large habitat areas using greenways and highway underpasses. It is based on an evaluation of: (1) spatial patterns and characteristics; (2) needs for local wildlife species; and (3) patterns of proposed ownership. The proposed habitat network comprises over 270,000 acres, and runs from the Santa Monica Mountains, which extend for 50 miles along the Pacific Ocean, to the Simi Hills, a smaller range linking the Santa Monica range to the Santa Susana Mountains, which connect to the mountains of Los Padres and Angeles National Forests further east. This vegetated area, once contiguous, is now dissected by a number of freeways and suburban developments that inhibit the movement of large animals, such as bobcat, mountain lion, and black bear. The habitat network has been identified and planned, based on such planning and design criteria as providing cover, water, habitat diversity, insulating corridors from human activity, nodes, and multiple pathways.

Reference:
Smith, D. S. 1993. "Greenway case studies." In Smith, D. S. and P.C. Hellmund, eds., *Ecology of Greenways. Design and Function of Linear Conservation Areas*. Univ. of Minn. Press, Minneapolis, pp. 161-208.

12. MESO SCALE. Configuration of deer habitats and rural housing developments (Montana, U.S.A.)

A study was undertaken to determine the effects of expanded housing developments on deer (i.e., white-tailed and mule) populations in Gallatin County, Montana, U.S.A. Background rationale for the study was based on a 50% increase in the number of rural residences over a 10-year period between 1970-80 in the county, with the amount of land in subdivisions increasing from 63 to 81 square kilometers. The study area comprised 1000 square kilometers, with elevations ranging from 1300 to 2000 meters. Specific study objectives were to: (1) determine the relative changes in populations and distribution of deer with increased housing; and (2) measure the impacts of housing density on movement and activity patterns of deer.

Results show an inverse relation between housing density and the number of deer observed. The home ranges of white-tailed deer decrease in size and become more linear as housing density increases. Of particular importance to deer movement and activity is the use of patches of cover which are either located in close proximity to each other and/or linked via travel corridors (i.e., stream or river banks).

Management implications for a strategy to benefit deer, based on these study results, would be to increase the density of housing on already developed areas, especially those of little value to wildlife and agriculture, rather than to develop new areas. In addition, for existing housing developments, provisions for leaving intact planting cover around housing clusters, and for not building within the band of vegetation along rivers and streams, would be beneficial to deer.

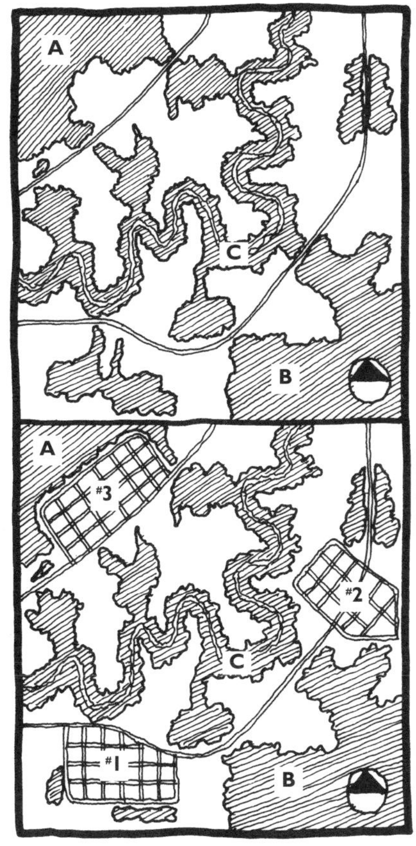

Housing developments and deer.

The adjacent illustration shows a hypothetical "*Before*" landscape situation (i.e., top half of map), as well as three "*After*" scenarios (i.e., bottom half), in which alternative housing developments #1, #2, and #3 are considered. In the "*Before*" case, primary deer habitat exists in large patches, A and B. Riparian corridor C largely connects these two deer habitats, although movement would be interrupted somewhat by the two roadways traversing the area. The remaining area of this landscape is in agricultural production.

Adapting the Gallatin County case study results to protect deer habitat, in the above hypothetical case would indicate that of the three development alternatives shown, #1 would be best, with #2 as an intermediate option, and #3 as the worst scenario.

Development #1 takes place within a previously isolated small habitat patch, which is neither an effective stepping stone from patch A to B, nor a large enough patch to consider protecting. Development #2, although not adding to further fragmentation of deer habitat, nor inducing any incremental obstacles to deer movement, is located on prime agricultural land. Development #3 not only reduces primary deer habitat, it further isolates this patch from other deer habitat.

Reference:
Vogel, W. O., 1989. "Response of deer to density and distribution of housing in Montana." *Wildlife Society Bulletin* 17(4), pp. 406-413.

13. MICRO SCALE. Agricultural drainage management techniques using riparian vegetation (North Carolina, U.S.A.)

Several studies in the Coastal Plain area of North Carolina have as their objective to observe the effect of riparian areas on the fate of nonpoint pollutants leaving agricultural fields, and to relate this to water quality downstream. Prior to this research, riparian areas were thought to represent opportunities for water treatment before surface drainage reached a perennial stream. Riparian vegetation had been reported to maintain: (1) the stability of the stream channel; and (2) the quality of water in both intermittent and continuous flow systems.

These research projects, which were conducted across four drainage basins, demonstrate the role riparian areas have in treating nitrogen, phosphorous, and sediment leaving cultivated fields. The effects of geomorphology and riparian vegetation on water movement, deposition-erosion, and nutrient dynamics cannot be easily separated from each other; there exists a complex interdependence.

These studies show how riparian areas act as chemical filters for nitrogen, treating agricultural run-off via subsurface flow through denitrification. Although the width of riparian vegetation needed for nitrogen treatment is not certain and depends on several factors, even narrow strips here provided noticeable protection. Floodplain swamps additionally proved effective in controlling sediment deposition and phosphorous uptake. The larger the floodplain swamp area, the greater the importance as a sediment trap and phosphorous sink.

References:

Cooper, J.R., J.W. Gilliam, and T.C. Jacobs, 1986. "Riparian areas as a control of nonpoint pollutants." *Journal Series of the N.C. Agricultural Research Service Paper* No. 10107, pp. 166-190.

Gilliam, J.W., R.W. Skaggs, and C.W. Doty, 1986. "Controlled agricultural drainage: an alternative to riparian vegetation." *N.C. Agricultural Research Service Paper* No. 10109, pp. 225-243.

14. MESO SCALE. Wildlife and water protection network: Pinhook Swamp and Suwannee River (Florida, U.S.A.)

Rampant urbanization, sprawl, and road construction in Florida has catalyzed public concern for species loss. A conservation goal has emerged to maintain connectivity among natural areas, thereby preserving key wildlife movement routes.

Two large, federally-owned habitats, the Okefenokee Swamp National Wildlife Refuge (approximately 400,000 acres) and Osceola National Forest (approximately 160,000 acres), are only miles apart. To prevent the strangling noose of development, linking these two large patches was identified as a top priority. A broad linkage of some 60,000 acres surrounding Pinhook Swamp was accomplished. This five-mile-wide wildlife movement corridor results in the creation of over 600,000 acres of contiguous habitat, home to many rare and endangered plant and animal species. As a result of this newly protected, larger area, individuals of numerous interior species can continue to move between the large patches of natural vegetation. And the total dumbbell-shaped area is considered sufficient to support viable populations of large-home-range species such as panthers and bears for long-term survival.

A proposed Suwannee River Corridor, which links the Osceola-Okefenokee-Pinhook system to the Gulf of Mexico, maintains connectivity along one of the last free-flowing rivers in the southeastern U.S. It also connects a diversity of habitats and a large collection of small protected patches. In the case of global warming and sea level rise, such a corridor would link coastal zones with interior habitat, thus enhancing long-term species migration and survival. Of the 426 miles of total river frontage, this proposal protects 152 miles for wildlife movement. Although as currently conceived, gaps would still exist, contingencies are provided to minimize or prevent disruptive activities along the entire length of the riparian corridor.

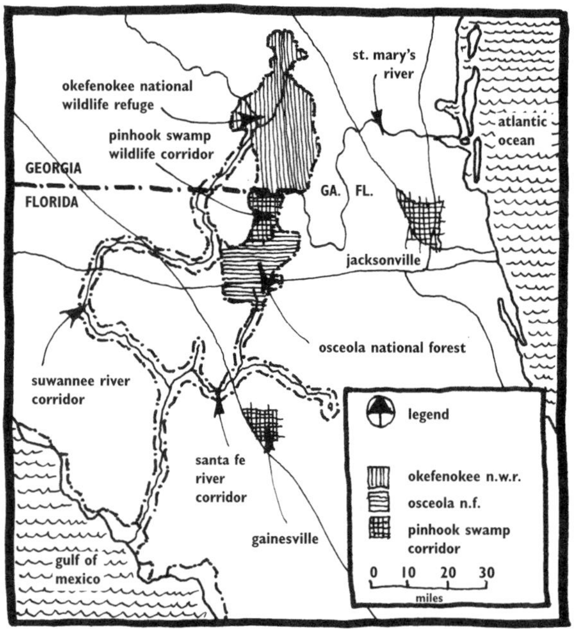

River network and Pinhook Swamp linkage.

Reference:

Smith, D.S. 1993. "Greenway case studies." In Smith, D.S. and P.C. Hellmund, eds., *Ecology of Greenways. Design and Function of Linear Conservation Areas*. Univ. of Minn. Press, Minneapolis, pp. 161-208.

SUMMARY AND CONCLUSION

Fifty-five principles or concepts of landscape ecology are presented. Almost all are stated in a single sentence, and are accompanied by a diagram illustrating the principle. The principles are grouped in four general categories: patches; edges/boundaries; corridors/connectivity; and mosaics. Dozens of references are given for each category, with key references highlighted at the end of each section.

In the patch category, principles focus on the ecological effects of size, number, and location of patches. The second category focuses on edge structure, plus boundary and patch shape. The third emphasizes connectivity provided by corridors and stepping stones, road and windbreak corridors, and stream and river corridors. Fourth, mosaics are represented by the ecological effects of networks, fragmentation, and scale.

Practical applications of landscape ecology principles are then illustrated. Schematic or conceptual landscape changes are applied in six cases over a broad range of scales, illustrating that the principles are effective independent of the size of a project. These applications state a problem or proposed change in the landscape, and portray "better" and "worse" designs, together with rationales.

Fourteen case studies that apply landscape ecology principles are outlined. The examples are of landscapes worldwide, and include a wide range of spatial scales. Projects focused on patches, edges/boundaries, corridors/connectivity, and mosaics are included, and offer lessons from both failures and successes.

Spatial pattern matters. It is no longer appropriate to plan based on totals or averages of prices, jobs, wages, parkland, bicycle paths, logging area, water flows, and so forth. Rather, the arrangement of land uses and habitats is crucial to planning, conservation, design, management, and policy.

Furthermore, context is usually more important than content. The attributes within a location are its usual descriptors. Yet the characteristics of the surrounding adjacent land-uses, of the upstream-upwind-upslope areas, and of the downstream-downwind-downslope areas are usually more important descriptors of the location. Sites are linked in a mosaic, where a change here affects many sites there.

The landscape ecology principles used in landscape architecture and land-use planning are here to stay. They are increasing in number and combinations among them are emerging. Some principles and patterns, such as vegetated corridors along major streams and a few large patches of natural vegetation, are "indispensable," i.e., no known or feasible alternative exists for providing their many major ecological benefits. Other disciplines are rapidly absorbing the landscape ecology principles, yet a distinct opportunity remains for land-use planners and landscape architects to "capture the principles and grab the future."

Some view land primarily as a source of wealth, a commodity that is bought and sold, an investment, a subject of laws and regulations, a matter of real estate, an object for tax policies, or a matter of economics. Others view land primarily as a living dynamic system, a place to live, a habitat containing plants and animals, a site of history, culture, aesthetics, and inspiration, or something that is planned, conserved, designed, managed, and cared for. Which perspective most motivates the reader? Which perspective is a basis for optimism about the future of society and nature? The principles in this book are usable now. They represent an embryo of optimism for the future.

ADDITIONAL REFERENCES

PATCHES

Abele, L.G. and E.F. Connor. 1979. "Application of island biogeographic theory to refuge design: making the right decisions for the wrong reasons." *Proceedings of Conference on Scientific Research in National Parks* 1, pp. 89-94.

Ambuel, B. and S.A. Temple. 1983. "Area dependent changes in the bird communities and vegetation of southern Wisconsin forests." *Ecology* 64, pp. 1057-1068.

Askins, R.A., M.J. Philbrick and D.S. Sugeno. 1987. "Relationship between the regional abundance of forest and the composition of forest bird communities." *Biological Conservation* 39, pp. 129-152.

Blake, J.G. and J.R. Karr. 1984. "Species composition of bird communities and the conservation benefit of large versus small forests." *Biological Conservation* 30, pp. 173-187.

Boecklen, W.J. 1986. "Optimal design of nature reserves: consequences of genetic drifts." *Biological Conservation* 38, pp. 328-338.

Boecklen, W.J., and N.J. Gotelli. 1984. "Island biogeographic theory and conservation practice: species-area or specious-area relationships?" *Biological Conservation* 29, pp. 63-80.

Brown, J.H. and A. Kodric-Brown. 1977. "Turnover rates in insular biogeography: effect of immigration on extinction." *Ecology* 58, pp. 445-449.

Burkey, T.V. 1989. "Extinction in nature reserves: the effects of fragmentation and the importance of movement between reserve fragments." *Oikos* 55, pp. 75-81.

Butcher, G.S., W.A. Niering, W.J. Barry and R.H. Goodwin. 1981. "Equilibrium biogeography and the size of nature preserves: an avian case study." *Oecologia* 49, pp. 29-37.

Connor, E.F., and E.D. McCoy. 1979. "The statistics and biology of the species-area relationship." *American Naturalist* 113, pp. 791-833.

Diamond, J.M. 1975. "The island dilemma: lessons of modern biogeographic studies for the design of natural reserves." *Biological Conservation* 7, pp. 129-146.

Diamond, J.M. 1984. " 'Normal' extinctions of isolated populations." In Nitecki, M.H., ed. *Extinctions*. University of Chicago Press, Chicago, pp. 191-246.

Fahrig, L. and G. Merriam. 1985. "Habitat patch connectivity and population survival." *Ecology* 66, pp. 1762-1768.

Forman, R.T.T. and R.E.J. Boerner. 1981. "Fire frequency and the Pine Barrens of New Jersey." *Bulletin of the Torrey Botanical Club* 108, pp. 34-50.

Forman, R.T.T. and M. Godron. 1986. *Landscape Ecology*. John Wiley, New York.

Franklin, J.F., and R.T.T. Forman. 1987. "Creating landscape patterns by cutting: ecological consequences and principles." *Landscape Ecology* 1, pp. 5-18.

Freemark, K.E., and G. Merriam. 1986. "Importance of area and habitat heterogeneity to bird assemblages in temperate forest fragments." *Biological Conservation* 36, pp. 115-141.

Galli, A.E., C.F. Leck and R.T.T. Forman. 1976. "Avian distribution patterns in forest islands of different sizes in central New Jersey." *Auk* 93, pp. 356-364.

Gilbert, F.S. 1980. "The equilibrium theory of island biogeography: fact or fiction?" *Journal of Biogeography* 7, pp. 209-235.

Gilpin, M.E., and J.A. Diamond. 1981. "Immigration and extinction probabilities for individual species." *Proceedings of the National Academy of Sciences (USA)*, pp. 392-396.

Goeden, G.B., 1979. "Biogeographic theory as a management tool." *Environmental Conservation* 6, pp. 27-32.

Gutzwiller, K.J., and S.H. Anderson, 1992. "Interception of moving organisms: influences of patch shape, size, and orientation on community structure." *Landscape Ecology* 6, pp. 293-303.

Henderson, M.T., G. Merriam, and J. Wegner. 1985. "Patchy environments and species survival: chipmunks in an agricultural mosaic." *Biological Conservation* 31, pp. 95-105.

Hobbs, E.R. 1988. "Species richness of urban forest patches and implications for urban landscape diversity." *Landscape Ecology* 1, pp. 141-152.

Janzen, D.H. 1983. "No park is an island: increase in interference from outside as park size decreases." *Oikos* 41, pp. 402-410.

Jennersten, O., J. Loman, A.P. Møller, J. Robertson, and B. Widen. 1992. "Conservation biology in agricultural habitat islands." In Hansson, L., ed. *Ecological Principles of Nature Conservation.* Elsevier Science Publishers, London, pp. 394-424.

Johnson, A.R., J.A. Wiens, B.T. Milne, and T.O. Crist. 1992. "Animal movements and population dynamics in heterogeneous landscapes." *Landscape Ecology* 7, pp. 63-75.

Landers, J.L., R.J. Hamilton, A.S. Johnson, and R.L. Marchington. 1979. "Foods and habitat of black bears in southeastern North Carolina." *Journal of Wildlife Management* 43, pp. 143-153.

Lynch, J.F., and D.A. Saunders. 1991. "Responses of bird species to habitat fragmentation in the wheatbelt of Western Australia: interiors, edges, and corridors." In Saunders, D.A., and R.J. Hobbs, eds. *Nature Conservation 2: The Role of Corridors.* Surrey Beatty, Chipping Norton, Australia. pp. 143-158.

MacDonald, D.W., and H. Smith. 1990. "Dispersal, dispersion and conservation in the agricultural ecosystem." In Bunce, R.G.H., and D.C. Howard, eds. *Species Dispersal in Agricultural Habitats.* Belhaven Press, London, pp. 18-64.

Middleton, J., and G. Merriam, 1986. "Woodland mice in a farmland mosaic." *Journal of Applied Ecology* 23, pp. 713-720.

Nilsson, S.G., and J. Bengtsson. 1988. "Habitat diversity or area Per Se? Species richness of woody plants, carabid beetles and land snails on islands." *Journal of Animal Ecology* 57, pp. 685-704.

Pease, C.M., R. Lande, and J.J. Bull. 1989. " A model of population growth, dispersal and evolution in a changing environment." *Ecology* 70, pp. 1657-1664.

Pickett, S.T.A., and J.N. Thompson. 1978. "Patch dynamics and the design of nature reserves." *Biological Conservation* 13, pp. 27-37.

Rafe, R.W., M.B. Usher, and R.G. Jefferson. 1985. "Birds on reserves: the influence of area and habitat on species richness." *Journal of Applied Ecology* 22, pp. 327-335.

Saunders, D.A. 1989. "Changes in the avifauna of a region, district, and remnant as a result of fragmentation of native vegetation: the wheatbelt of Western Australia: A case study." *Biological Conservation* 50, pp. 99-135.

Saunders, D.A., and J.A. Ingram. 1987. "Factors affecting survival of breeding populations of Carnaby's cockatoo *Calyptorhynchus funereus latirostris* in remnants of native vegetation." In Saunders, D.A., G.W. Arnold, A.W. Burbidge, and A.J.M. Hopkins, eds. *Nature Conservation: The Role of Remnants of Native Vegetation.* Surrey Beatty, Chipping Norton, Australia, pp. 249-258.

Taylor, A.D. 1990. "Metapopulation dispersal, and predator-prey dynamics: an overview." *Ecology* 71, pp. 429-436.

Taylor, A.D. 1991. "Studying metapopulation effects in predator-prey systems." *Biological Journal of the Linnean Society* 42, pp. 305-323.

Turner, M.G. 1989. "Landscape ecology: the effect of pattern on process." *Annual Review of Ecology and Systematics* 20, pp. 171-197.

Wegner, J., and G. Merriam. 1979. "Movements by birds and small mammals between a wood and adjoining farmland habitats." *Journal of Applied Ecology* 16, pp. 349-358.

EDGES AND BOUNDARIES

Chasko, G.G. and J.E. Gates. 1982. "Avian habitat suitability along a transmission-line corridor in an oak-hickory forest region." *Wildlife Monographs* 82, pp. 1-41.

De Walle, D.R. 1983. "Wind damage around clearcuts in the ridge and valley province of Pennsylvania." *Journal of Forestry* 81, pp. 158-159.

Feder, J. 1988. *Fractals*. Plenum Press, New York.

Forman, R.T.T. and P.N. Moore. 1992. "Theoretical foundations for understanding boundaries in landscape mosaics." In Hansen, A.J. and F. di Castri, eds. *Landscape Boundaries: Consequences for Biotic Diversity and Ecological Flows*. Springer-Verlag, New York, pp. 236-258.

Gardner, R.H., R.V. O'Neill, M.G. Turner and V.H. Dale. 1989. "Quantifying scale-dependent effects of animal movement with simple percolation models." *Landscape Ecology* 3, pp. 217-227.

Gardner, R.H., M.G. Turner, V.H. Dale and R.V. O'Neill. 1992. "A percolation model of ecological flows." In Hansen, A.J. and F. di Castri, eds. *Landscape Boundaries: Consequences for Biotic Diversity and Ecological Flows*. Springer-Verlag, New York, pp. 259-269.

Gates, J.E. and L.W. Gysel. 1978. "Avian nest dispersion and fledgling success in field-forest ecotones." *Ecology* 59, pp. 871-883.

Hanley, T.A. 1983. "Black-tailed deer, elk, and forest edge in a western Cascades watershed." *Journal of Wildlife Management* 47, pp. 237-242.

Johnson, A.R., J.A. Wiens and B.T. Milne. 1992. "Animal movements and population dynamics in heterogeneous landscapes." *Landscape Ecology* 7, pp. 63-75.

Kareiva, P. 1982. "Experimental and mathematical analyses of herbivore movement: quantifying the influence of plant spacing and quality on foraging discrimination." *Ecological Monographs* 52, pp. 261-282.

Kroodsma, R.L. 1982. "Bird community ecology on power-line corridors in East Tennessee." *Biological Conservation* 23, pp. 79-94.

Leopold, A.S. 1933. *Game Management*. Charles Schribner's Sons, New York.

McCreary, D.D. and D.A. Perry. 1983. "Strip thinning and selective thinning in Douglas-fir." *Journal of Forestry* 81, pp. 375-377.

Milne, B.T. 1991. "Lessons from applying fractal models to landscape patterns." In Turner, M.G. and R.H. Gardner, eds. *Quantitative Methods in Landscape Ecology*. Springer-Verlag, New York, pp. 199-235.

Milne, B.T. 1991. "Heterogeneity as a multiscale characteristic of landscapes." In Kolasa, J.and S.T.A. Pickett, eds. *Ecological Heterogeneity*. Springer-Verlag, New York.

Odum, E.P. and M.G. Turner. 1990. "The Georgia landscape: a changing resource." In Zonneveld, I.S. and R.T.T. Forman, eds. *Changing Landscapes: An Ecological Perspective*. Springer-Verlag, New York, pp. 137-164.

O'Neill, R.V., J.R. Krummel, R.H. Gardner, G. Sugihara, B. Jackson, D.L. DeAngelis, B.T. Milne, M.G. Turner, B. Zygmunt, S.W. Christensen, V.H. Dale and R.L. Graham. 1988. "Indices of landscape pattern." *Landscape Ecology* 1, pp. 153-162.

Orians, G.H. and N.E. Pearson. 1978. "On the theory of central place foraging." In Horns, D.J., G.R. Stairs and R.D. Mitchell, eds. *Analysis of Ecological Systems*. Ohio State University Press, Columbus, Ohio, pp. 155-177.

Rapoport, E.H. 1982. *Areography: Geographical Strategies of Species*. Fundacion Bariloche Series Number 1. Pergamon Press, New York.

Turner, M.G., R.H. Gardner, V.H. Dale and R.V. O'Neill. 1989. "Predicting the spread of disturbance across heterogeneous landscapes." *Oikos* 55, pp. 121-129.

van Leeuwen, C.G. 1981. "From ecosystem to ecodevice." In Tjallingii, S. and A. A. de Veer, eds. *Perspectives in Landscape Ecology*, Pudoc, Wageningen, Netherlands, pp. 29-34.

Wales, B.A. 1972. "Vegetation analysis of northern and southern edges in a mature oak-hickory forest." *Ecological Monographs* 42, pp. 451-471.

Wiens, J.A. 1976. "Population responses to patchy environments." *Annual Review of Ecology and Systematics* 7, pp. 81-120.

Wiens, J.A. and B.T. Milne. 1989. "Scaling of 'landscapes' in landscape ecology, or, landscape ecology from a beetle's perspective." *Landscape Ecology* 3, pp. 87-96.

Wilmanns, O. and J. Brun-Hool. 1982. "Irish mantel and saum vegetation." *Journal of Life Sciences, Royal Dublin Society* 3, pp. 165-174.

CORRIDORS AND CONNECTIVITY

Adams, L.W. and L.E. Dove. 1989. *Wildlife Reserves and Corridors in the Urban Environment*. National Institute for Urban Wildlife, Columbia, Maryland.

Adams, L.W. and A.D. Geis. 1983. "Effect of roads on small mammals." *Journal of Applied Ecology* 20, pp. 403-415.

Ambuel, B. and S.A. Temple. 1983. "Area dependent changes in the bird communities and vegetation of southern Wisconsin forests." *Ecology* 64, pp. 1057-1068.

Andrews, J. 1993. "The reality and management of wildlife corridors." *British Wildlife* 5, pp. 1-7.

Arnold, G.W. 1983. "The influence of ditch and hedgerow structure, length of hedgerows, and area of woodland and garden on bird numbers in farmland." *Journal of Applied Ecology* 20 pp. 731-750.

Baudry, J. 1984. "Effects of landscape structure on biological communities: the case of hedgerow network landscapes." In Brandt, J. and P. Agger, eds. *Proceedings of the First International Seminar on Methodology in Landscape Ecological Research and Planning*, Vol. 1, Roskilde Universitetsforlag, Roskilde, Denmark, pp. 55-65.

Baudry, J. 1988. "Hedgerows and hedgerow networks as wildlife habitat in agricultural landscapes." In Park, J.R., ed. *Environmental Management in Agriculture. European Perspectives*. Belhaven Press, London, pp. 111-124.

Baudry, J. and G. Merriam. 1988. "Connectivity and connectedness: functional versus structural patterns in landscapes." *Munsterische Geographische Arbeiten* 29, pp. 23-28.

Bennett, A.F. 1990. "Habitat corridors and the conservation of small mammals in a fragmented forest environment." *Landscape Ecology* 4, pp. 109-122.

Bennett, A.F. 1991. "What types of organisms will use corridors?" In Saunders, D.A. and R.J. Hobbs, eds. *Nature Conservation 2: The Role of Corridors*. Surrey Beatty, Chipping Norton, Australia, pp. 407-408.

Boone, G.S. and R. Tincklin. 1988. "The importance of hedgerow structure in the occurrence and density of small mammals." *Aspects of Applied Biology* 16, pp. 73-78.

Brockie, R.E., L.L. Loope, M.B. Usher and O. Hamann, 1988. "Biological invasions of island nature reserves." *Biological Conservation* 44, pp. 9-36.

Burel, F. and J. Baudry. 1990. "Hedgerow networks as habitats for forest species: implications for colonizing abandoned agricultural land." In Bunce, R.G.H. and D.C. Howard, eds. *Species Dispersal in Agricultural Habitats*. Belhaven Press, London, pp. 18-64.

Dawson, D. 1994. "Are habitat corridors conduits for animals and plants in a fragmented landscape? A review of the scientific evidence." *English Nature Research Report* 94, pp. 6-67.

Faaborg, J. 1979. "Qualitative patterns of avian extinction on neotropical land-bridge islands: lessons for conservation." *Journal of Applied Ecology* 16, pp. 99-107.

Fahrig, L. and G. Merriam. 1985. "Habitat patch connectivity and population survival." *Ecology* 66, pp. 1762-1768.

Forman, R.T.T. 1991. "Landscape corridors: from theoretical foundations to public policy." In Saunders, D.A. and R.J. Hobbs, eds. *Nature Conservation 2: The Role of Corridors*. Surrey Beatty, Chipping Norton, Australia, pp. 71-84.

Forman, R.T.T. 1984. "Hedgerows and hedgerow networks in landscape ecology." *Environmental Management* 8, pp. 495-510.

Friend, G.R. 1991. "Does corridor width and composition affect movement?" In Saunders, D.A. and R.J. Hobbs, eds. *Nature Conservation 2: The Role of Corridors*. Surrey Beatty, Chipping Norton, Australia, pp. 404-405.

Gilpin, M.E. 1980. "The role of stepping stone islands." *Theoretical Population Biology* 17, pp. 247-253.

Gilpin, M.E. and J.A. Diamond. 1981. "Immigration and extinction probabilities for individual species. Relation to incidence..." *Proceedings of the National Academy of Sciences (USA)* 78, pp. 392-396.

Goldstein-Golding, E.L. 1991. "The ecology and structure of urban green spaces." In Bell, S.S., E.D. McCoy and H.R. Muushinsky, eds. *Habitat Structure. The Physical Arrangement of Objects in Space*. Chapman and Hall, London, pp. 392-411.

Hanski, I. 1982. "On temporal and spatial variation in animal populations." *Acta Zoologica Fennici* 19, pp. 21-37.

Hanski, I. 1991. "Single-species metapopulation dynamics: concepts, models and observations." *Biological Journal of the Linnean Society* 42, pp. 17-38.

Hill, M.O. and P.D. Carey. 1994. "The Role of Corridors, Stepping Stones and Islands for Species Conservation in a Changing Climate." *English Nature Report* 75.

Hobbs, R.J. 1992. "The role of corridors in conservation: solution or bandwagon?" *Trends in Ecology and Evolution* 7, pp. 389-392.

Hobbs, R.J. and A.J.M. Hopkins. 1991. "The role of conservation corridors in a changing climate." In Saunders, D.A. and R.J. Hobbs, eds. *Nature Conservation 2: The Role of Corridors*. Surrey Beatty, Chipping Norton, Australia, pp. 281-290.

Johnson, R.J. and M.M. Beck. 1988. "Influences of shelterbelts on wildlife management and biology." *Agriculture, Ecosystems and Environment* 22/23, pp. 301-305. (Reprinted 1988 in *Windbreak Technology*. Elsevier, Amsterdam).

Lyle, J. and R.D. Quinn. 1991. "Ecological corridors in urban southern California." In *Wildlife Conservation in Metropolitan Environments*, National Institute for Urban Wildlife, Columbia, Maryland.

Mader, H.-J. 1984. "Animal habitat isolation by roads and agricultural fields." *Biological Conservation* 29, pp. 81-96.

Mader, H.-J., C. Schell and P. Kornacker. 1990. "Linear barriers to arthropod movements in the landscape." *Biological Conservation* 54, pp. 115-128.

Merriam, G. 1984. "Connectivity: a fundamental ecological characteristic of landscape pattern." In Brandt, J. and P. Agger, eds. *Methodology in Landscape Ecological Research and Planning*, Vol. 1. Roskilde Universitetsforlag GeoRuc, Roskilde, Denmark, pp. 5-16.

Merriam, G.,1991. "Corridors and connectivity: animal populations in heterogeneous environments." In Saunders, D.A. and R.J. Hobbs, eds. *Nature Conservation 2: The Role of Corridors.* Surrey Beatty, Chipping Norton, Australia, pp. 133-142.

Merriam, G., M. Kozakiewicz, E. Tsuchiya and K. Hawley. 1989. "Barriers as boundaries for metapopulations and demes of *Peromyscus leucopus* in farm landscapes." *Landscape Ecology* 2, pp. 227-235.

Merriam, G. and A. Lanoue. 1990. "Corridor use by small mammals: field measurement for three experimental types of *Peromyscus leucopus.*" *Landscape Ecology* 4, pp. 123-131.

Munguia, M.L. and J.A. Thomas. 1992. "Use of road verges by butterfly and burnet populations, and the effect of roads on adult dispersal and mortality." *Journal of Applied Ecology* 29, pp. 316-329.

Nicholls, A.O. and C.R. Margules. 1991. "The design of studies to demonstrate the biological importance of corridors." In Saunders, D.A. and R.J. Hobbs, eds. *Nature Conservation 2: The Role of Corridors.* Surrey Beatty, Chipping Norton, Australia, pp. 49-61.

Noss, R.F. 1983. "A regional landscape approach to maintain diversity." *Bioscience* 33, pp. 700-706.

Noss, R.F. 1987. "Corridors in real landscapes: a reply to Simberloff and Cox." *Conservation Biology* 1, pp. 159-164.

Noss, R.F. and L. Harris. 1986. "Nodes, networks, and MUMs: preserving diversity at all scales." *Environmental Management* 10, pp. 299-309.

Opdam, P. 1990. "Dispersal in fragmented populations: the key to survival." In Bunce, R.G.H. and D.C. Howard, eds. *Species Dispersal in Agricultural Habitats.* Belhaven Press, London, pp. 3-17.

Ouellet, H. 1967. "Dispersal of land birds on the Islands of the Gulf of St. Lawrence." *Canadian Journal of Zoology* 45, pp. 1149-1167.

Panetta, F.D. and A.J.M. Hopkins. 1991. "Weeds in corridors: invasion and management." In Saunders, D.A. and R.J. Hobbs, eds. *Nature Conservation 2: The Role of Corridors.* Surrey Beatty, Chipping Norton, Australia, pp. 341-351.

Saunders, D.A. and J.A. Ingram. 1987. "Factors affecting survival of breeding populations of Carnaby's cockatoo *Calyptorhynchus funereus latirostris* in remnants of native vegetation." In Saunders, D.A., G.W. Arnold, A.W. Burbidge and A.J.M. Hopkins, eds. *Nature Conservation: The Role of Remnants of Native Vegetation.* Surrey Beatty, Chipping Norton, Australia, pp. 249-258.

Saunders, D.A. and R.J. Hobbs. 1989. "Corridors for conservation." *New Scientist* 28, pp. 63-68.

Saunders, D.A. and C.P. de Rebeira. 1991. "Values of corridors to avian populations in a fragmented landscape." In Saunders, D.A. and R.J. Hobbs, eds. *Nature Conservation 2: The Role of Corridors.* Surrey Beatty, Chipping Norton, Australia, pp. 221-240.

Simberloff, D.S. and J. Cox. 1987. "Consequences and costs of conservation corridors." *Conservation Biology* 1, pp. 63-71.

Spellerberg, I.F. and M. Gaywood. 1993. "Linear features: linear habitats and wildlife corridors." *English Nature Report.*

van der Zande, A.N., W.J. ter Keurs and W.J. van der Weijden. 1980. "The impact of roads on the densities of four bird species in an open field habitat - evidence of a long distance effect." *Biological Conservation* 18, pp. 299-321.

Verkaar, H.J. 1990. "Corridors as a tool for plant species conservation?" In Bunce, R.G.H. and D.C. Howard, eds. *Species Dispersal in Agricultural Habitats.* Belhaven Press, London, pp. 82-97.

MOSAICS

Adams, L.W. and L.E. Dove. 1989. *Wildlife Reserves and Corridors in the Urban Environment.* National Institute for Urban Wildlife, Columbia, Maryland.

Ambuel, B. and S.A. Temple. 1983. "Area dependent changes in the bird communities and vegetation of southern Wisconsin forests." *Ecology* 64, pp. 1057-1068.

Askins, R.A., M.J. Philbrick and D.S. Sugeno. 1987. "Relationship between the regional abundance of forest and the composition of forest bird communities." *Biological Conservation* 39, pp. 129-152.

Barloy, J. 1980. "Consequences sur la production végétale agricole de l'amenagement du bocage dans l'Ouest de la France." *Bulletin Technologique Information*, pp. 353-355.

Baudry, J. 1984. "Effects of landscape structure on biological communities: the case of hedgerow network landscapes." In Brandt, J. and P. Agger, eds. *Proceedings of the First International Seminar on Methodology in Landscape Ecological Research and Planning*, Vol. 1. Roskilde Universitetsforlag, Roskilde, Denmark, pp. 55-65.

Baudry, J. 1988. "Hedgerows and hedgerow networks as wildlife habitat in agricultural landscapes." In Park, J.R., ed. *Environmental Management in Agriculture.* European Perspectives. Belhaven Press, London, pp. 111-124.

Bennett, A.F. 1990. "Habitat corridors and the conservation of small mammals in a fragmented forest environment." *Landscape Ecology* 4, pp. 109-122.

Burel, F. and J. Baudry. 1990. "Hedgerow networks as habitats for forest species: implications for colonizing abandoned agricultural land." In Bunce, R.G.H. and D.C. Howard, eds. *Species Dispersal in Agricultural Habitats.* Belhaven Press, London, pp. 18-64.

Brocke, R.H., J.P. O'Pezio and K.A. Gustafson. 1990. "A forest management scheme mitigating impact of road networks on sensitive wildlife species." In *Is Forest Fragmentation a Management Issue in the Northeast?* General Technical Report NE-140, USDA Forest Service, Radnor, Pennsylvania, pp. 13-17.

Bryer, J.B. 1983. "The effects of a geometric redefinition of the classical road and landing spacing model through shifting." *Forest Science* 29, pp. 670-674.

Constant, P., M.C. Eybert and R. Maheo. 1976. "Aviafune reproductrice du bocage de l'Ouest." In *Les Bocages: Histoire, Ecologie, Economie*, Institut National de la Recherche Agronomique, Centre National de la Recherche Scientifique, et Université de Rennes, Rennes, France, pp. 327-332.

Fahrig, L. and G. Merriam. 1985. "Habitat patch connectivity and population survival." *Ecology* 66, pp. 1762-1768.

Forman, R.T.T. 1987. "The ethics of isolation, the spread of disturbance, and landscape ecology." In Turner, M.G., ed. *Landscape Heterogeneity and Disturbance.* Springer-Verlag, New York, pp. 213-229.

Forman, R.T.T. 1990. "Ecologically sustainable landscapes: the role of spatial configuration." In Zonneveld, I.S. and R.T.T. Forman, eds. *Changing Landscapes: An Ecological Perspective.* Springer-Verlag, New York, pp. 261-278.

Forman, R.T.T. 1994. "A wildlife test of design and planning." *Studio Works 2*, Harvard University Graduate School of Design, Cambridge, p. 69.

Forman, R.T.T. and S.K. Collinge. 1995. "The 'spatial solution' to conserving biodiversity in landscapes and regions." In DeGraaf, R.M. and R.I. Miller, eds. *Conservation of Faunal Diversity in Forested Landscapes.* Chapman and Hall, London, in press.

Gardner, R.H., M.G. Turner, V.H. Dale and R.V. O'Neill. 1992. "A percolation model of ecological flows." In Hansen, A.J. and F. di Castri, eds. *Landscape Boundaries: Consequences for Biotic Diversity and Ecological Flows*, Springer-Verlag, New York, pp. 259-269.

Guyot, G. and M. Verbrugghe. 1976. "Influence du bocage sur le climat d'une petite région." In *Les Bocages: Histoire, Ecologie, Economie*. Institut National de la Recherche Agronomique, Centre National de la Recherche Scientifique, et Université de Rennes, Rennes, France, pp. 131-136.

Hanley, T.A. 1983. "Black-tailed deer, elk, and forest edge in a western Cascades watershed." *Journal of Wildlife Management* 47, pp. 237-242.

Hansen, A.J., S.L. Garman and B. Marks. 1993. "An approach for managing vertebrate diversity across multiple-use landscapes." *Ecological Applications* 3, pp. 481-496.

Harris, L.D. 1984. *The Fragmented Forest: Island Biogeography Theory and the Preservation of Biotic Diversity*. University of Chicago Press, Chicago.

Henderson, M.T., G. Merriam and J. Wegner. 1985. "Patchy environments and species survival: chipmunks in an agricultural mosaic." *Biological Conservation* 31, pp. 95-105.

Howe, R.W. 1984. "Local dynamics of bird assemblages in small forest habitat islands in Australia and North America." *Ecology* 65, pp. 1585-1601.

Johnson, A.R., J.A. Wiens, B.T. Milne and T.O. Crist. 1992. "Animal movements and population dynamics in heterogeneous landscapes." *Landscape Ecology* 7, pp. 63-75.

Lidicker, W.Z., J.O. Wolff, L.N. Lidicker and M.H. Smith. 1992. "Utilization of a habitat mosaic by cotton rats during a population decline." *Landscape Ecology* 6, pp. 259-268.

Lord, J.M. and D.A. Norton. 1990. "Scale and the spatial concept of fragmentation." *Conservation Biology* 4, pp. 197-202.

Lynch, J.F. and D.F. Whigham. 1984. "Effects of forest fragmentation on breeding bird communities in Maryland, USA." *Biological Conservation* 28, pp. 287-324.

Lyon, L.J. 1979. "Habitat effectiveness for elk as influenced by roads and cover." *Journal of Forestry* 77, pp. 658-660.

Margules, C.R. and A.O. Nicholls. 1987. "Assessing the conservation value of remnant habitat 'islands': mallee patches on the western Eyre Peninsula, South Australia." In Saunders, D.A., G.W. Arnold, A.A. Burbidge and A.J.M. Hopkins, eds. *Nature Conservation: The Role of Remnants of Native Vegetation*. Surrey Beatty, Chipping Norton, Australia, pp. 89-102.

Middleton, J., and G. Merriam. 1981. "Woodland mice in a farmland mosaic." *Journal of Applied Ecology* 18, pp. 703-710.

Morris, M.G. and K.H. Lokhani. 1979. "Responses of grassland invertebrates to management by cutting." *Journal of Applied Ecology* 16, pp. 77-98.

Murphy, D.R., K. Freas and S. Weiss. 1990. "An environment-metapopulation approach to population viability analysis for a threatened invertebrate." *Conservation Biology* 4, pp. 41-51.

Noss, R.F. 1983. "A regional landscape approach to maintain diversity." *BioScience* 33, pp. 700-706.

O'Neill, R.V., D.L. DeAngelis, J.B. Waide and T.F.H. Allen. 1986. *A Hierarchical Concept of Ecosystems*. Princeton University Press, Princeton, New Jersey.

Peterken, G.F., D. Ausherman, M. Buchenau and R.T.T. Forman. 1992. Old-growth conservation within British upland conifer plantations. *Forestry* 65, pp. 127-144.

Pollard, E., M.D. Hooper and N.W. Moore. 1974. *Hedges*. Collins, London.

Sharpe, D.M., F.W. Stearns, R.L. Burgess and W.C. Johnson. 1981. "Spatio-temporal patterns of forest ecosystems in man-dominated landscapes of the eastern United States." In Tjallingii, S.P. and A.A. de Veer, eds. *Perspectives in Landscape Ecology*, Pudoc, Wageningen, Netherlands, pp. 109-116.

Stritch, L.R. 1990. "Landscape-scale restoration of barrens-woodland within the oak-hickory forest mosaic." *Restoration and Management News* 8, pp. 73-77.

Swanson, F.J., J.F. Franklin and J.R. Sedell. 1990. "Landscape patterns, disturbance, and management in the Pacific Northwest, USA." In Zonneveld, I.S. and R.T.T. Forman, eds. *Changing Landscapes: An Ecological Perspective*. Springer-Verlag, New York, pp. 191-213.

Turner, M.G. 1989. Landscape ecology: the effect of pattern on process. *Annual Review of Ecology and Systematics* 20, pp. 171-197.

Usher, M.B. 1987. "Effects of fragmentation on communities and populations: actions, reactions, and applications to wildlife conservation." In Saunders, D.A., G.W. Arnold, A.A. Burbidge and A.J.M. Hopkins. *Nature Conservation: The Role of Remnants of Native Vegetation*. Surrey Beatty, Chipping Norton, Australia, pp. 103-121.

Verboom, J. and R. van Apeldoorn. 1990. "Effects of fragmentation on the red squirrel, *Sciurus vulgaris* L." *Landscape Ecology* 4, pp. 171-176.

Wales, B.A. 1972. "Vegetation analysis of northern and southern edges in a mature oak-hickory forest." *Ecological Monographs* 42, pp. 451-471.

Wegner, J. and G. Merriam. 1979. "Movements by birds and small mammals between a wood and adjoining farmland habitats." *Journal of Applied Ecology* 16, pp. 349-358.

Wilcox, B.A. and D.D. Murphy. 1985. "Conservation strategy: the effects of fragmentation on extinction." *American Naturalist* 125, pp. 879-887.

Yapp, W.B. 1973. "Ecological evaluation of a linear landscape." *Biological Conservation* 5, pp. 45-47.

Wenche E. Dramstad works primarily in landscape ecological design and management issues involved with agricultural landscapes in Norway. She received her Master of Nature Conservation from the Department of Biology and Nature Conservation at the Agricultural University of Norway in 1990. She is now completing her PhD thesis in landscape ecology at the same institution. Her particular interest is in landscape ecology as a bridge between ecologists and land use planners, while her PhD study focuses on insect behavior in agricultural landscapes.

James D. Olson is principal and co-founder of a recently established landscape and architectural design firm, specializing in landscape ecologically-based designs and management projects. He received his Master of Landscape Architecture from the Harvard University Graduate School of Design in 1995 and a Master of Business Administration from Columbia University Graduate Business School in 1983.

Richard T.T. Forman is the PAES Professor of Landscape Ecology at Harvard University. He has served as vice-president of the Ecological Society of America and the International Association of Landscape Ecology, and as President of the Torrey Botanical Club. He received an honorary Doctor of Human Letters from Miami University, a Fulbright scholarship in Columbia, a CNRS Chercheur in France, and the Lindback Foundation Award for Excellence in Teaching. He is a Fellow of the AAAS and of Clare Hall, Cambridge. Forman has published numerous scholarly articles in addition to six books.

Director of Lectures, Exhibitions, and Academic Publications
Brooke Hodge

Design
Nicole Juen Studio
Providence, Rhode Island

Production Coordination
Susan McNally

Line and Color Illustrations
Wenche E. Dramstad
James D. Olson
Richard T.T. Forman

Printing
Reynolds DeWalt Printing, Inc.
New Bedford, Massachusetts